Theory of Hydraulic Models

Macmillan Civil Engineering Hydraulics

General Editor: Professor E. M. Wilson, Ph.D., M.Sc., F.I.C.E.
Professor of Hydraulic Engineering, University of Salford

FLOW IN CHANNELS: *R. H. J. Sellin*
ENGINEERING HYDROLOGY: *E. M. Wilson*
COASTAL HYDRAULICS: *A. M. Muir Wood*
ANALYSIS OF SURGE: *John Pickford*
THEORY OF GROUNDWATER FLOW: *A. Verruijt*

Theory of Hydraulic Models

M. SELIM YALIN, M.A.S.C.E., M.I.C.E., M.E.I.C.
Queen's University at Kingston, Canada

MACMILLAN

First published 1971

Published by
THE MACMILLAN PRESS LTD
London and Basingstoke
Associated companies in New York, Toronto Dublin, Melbourne, Johannesburg and Madras

Printed in Great Britain at the
PITMAN PRESS
Bath

Preface

The present book concerns the design of hydraulic models. Theory cannot cover all the complications that are encountered in practice, so that almost every major project in the field of hydraulic engineering undergoes a 'model test' where, on a small scale model, the flows and their consequences corresponding to various versions of the scheme may be observed, and the relevant quantities measured. As a result of these observations and measurements, the most effective and rational design of the scheme can be determined.

A hydraulic model is a precision device for the experimental investigation of a hydromechanical phenomenon, which can give reliable information only if its scales are determined according to certain definite rules; that is, if it is designed correctly. If the design is not correct, then the model is wrong in principle. In that case, the employment of the most sophisticated instrumentation and measurement methods can only help to increase the accuracy of the wrong predictions. A small scale reproduction of a physical phenomenon can be a scientifically valid model only if a certain set of its measurable characteristics are related to their counterparts in the actual phenomenon, or prototype, by certain constant proportions which satisfy definite mathematical conditions. These constant proportions are referred to as scales, whereas the mathematical conditions which must be satisfied by the scales are called the criteria of similarity. It follows that the design and realization of a true model of a phenomenon can only be achieved if the similarity criteria of that phenomenon are known. These can be revealed by the mathematical relationships (usually differential equations of motion) describing the physical nature of the phenomenon under investigation. However, the accuracy, and thus the reliability of the results determined by this method, depend entirely on the reliability of the mathematical relationships used, and if the phenomenon has not been formulated mathematically then its criteria of similarity simply cannot be determined. This is an ironic situation, for a model is of greatest use in those cases which cannot be formulated theoretically.

Another, and considerably more effective way of determining the criteria of similarity is the dimensional method. Here, the criteria of similarity are supplied from dimensional study of the characteristics themselves and not from the mathematical relationships (containing the characteristics involved). In other words, the criteria of similarity are supplied by the dimensional method without the risk of misinterpretation of the physical nature of the phenomena (which might be inherent in mathematical relationships). The dimensional method has been the subject of considerable controversy and it has taken a long time for it to be accepted that it is not a 'dangerous tool which can often give erroneous results', but that it is a tool which, like every other mathematical tool, always gives correct results if correctly used and which gives erroneous results if erroneously used. After the appearance of the works of Bridgeman, Birkhoff, Langhaar, Comolet, Sedov, and others, in the last three decades, it has also been generally accepted that the theory of dimensions is not a set of trivial rules with no other use than to determine the exponents of a power product, but that it is what may be called the 'theory of experimental research'.

The application of the criteria of similarity to the design of hydraulic models is, in practice, frequently impeded by serious difficulties of a physical or technical nature. The fact that the prototype and its model are on the same planet implies that they are both subject to the same force of gravity, or that the scale of the acceleration due to gravity g is necessarily equal to unity. Similarly, the utilization of the prototype fluid (water) in the model means that the scales of fluid density and viscosity are also equal to unity. It is often very difficult, and sometimes impossible, to satisfy the criteria of similarity by using such 'unadjustable' values of the characteristics mentioned. In such cases the model cannot be designed according to the exact principles of the theory of models, and it is up to the knowledge, experience, and even intuition, of the experimenter to decide what would be the 'next best thing' to do in order to design a reliable model. These drawbacks do not, however, mean that the derivation of exact criteria of similarity is merely an academic exercise and thus that it might be possible to design a model without having any information on the principles of the theory of models (see Chapter 2). One can never know for a particular similarity problem how nearly an adopted practicable solution approximates the correct solution if the latter is unknown. Therefore, whether applicable or not, the exact solution should always be first determined, and then, if it proves to be impracticable, used as a frame of reference to judge the degree of correctness of the adopted solution.

I would like to express my thanks to Professor E. M. Wilson, the Editor of this Series, for his comments which have improved the text considerably. I am also grateful to Mr. C. H. Russell, the Director of the Hydraulics Research Station, Wallingford, for reviewing the manuscript and offering a number of valuable suggestions. My appreciation is extended to my colleagues, Professor A. W. Peterson and Dr J. W. Kamphuis, who were so helpful during the preparation of this work. Finally, I would like to thank my wife Cherrilyn for typing this difficult manuscript.

KINGSTON, Ontario M. S. YALIN
May 1970

Contents

List of Plates

1 Principles of the Theory of Dimensions

1.1 Dimensional and Dimensionless Quantities

The realm of mechanics consists of a variety of concepts such as energy, force, velocity, density, and so on. In the present book, the word 'mechanics' refers to classical or non-relativistic mechanics. No limit can be imposed upon the nature and the number of concepts; the progress of science means the birth of new ideas: the introduction of new concepts. On the other hand, each one of this unlimited number of concepts can be defined by means of only three independent entities—length, time and mass—referred to as fundamental entities. It is interesting to note that these three fundamental entities cannot themselves be defined. Nothing in the physical or external world is more obvious to us than what is implied by near and far, earlier and later, and lighter and heavier. We learn these notions, without being taught, simply by living in this world. A concept can be defined only by means of more familiar concepts, yet the forementioned terms remain obvious to us. Any entity that can be measured and expressed in numbers is a quantity. It follows that various mechanical quantities can be regarded merely as compositions of the same three measurable entities: length, time and mass. Let L, T and M be the units for length, time and mass. Since a mechanical quantity a can be considered as a composition of length, time and mass, the unit of a, denoted by $[a]$, must be a function of the fundamental units, i.e.

$$[a] = f(L, T, M) \tag{1.1}$$

The ratio of two different numerical values of a quantity a cannot depend on the choice of the fundamental units. One can prove that this physical requirement can be satisfied only if the function f has the form of a power product[1,2,3,4,5]. Accordingly, the unit $[a]$ of any quantity a is given by

$$[a] = L^{\alpha}T^{\beta}M^{\gamma} \tag{1.2}$$

where the nature of the quantity a is reflected by the numerical values of the exponents α, β and γ. For example, if a is a velocity $\alpha = 1$, $\beta = -1$ and $\gamma = 0$; if a is a force $\alpha = 1$, $\beta = -2$ and $\gamma = 1$, etc. The expression of the unit of a quantity a in terms of the fundamental units L, T and M (in other words the right hand side of eqn. (1.2)) is called *the dimension formula*, or simply *the dimension* of the quantity a. The quantity a is said to possess a dimension or to be a *dimensional quantity* if at least one of the the exponents α, β and γ is not zero. Any dimensional quantity which is not one of the fundamental quantities is usually referred to as a 'derived quantity'.

A dimensional quantity a encountered in mechanics is said to be

(i) a geometric quantity if $\alpha \neq 0;\ \beta = 0;\ \gamma = 0$
(ii) a kinematic quantity if $\alpha \neq 0;\ \beta \neq 0;\ \gamma = 0$
(iii) a dynamic quantity if $\alpha \neq 0;\ \beta \neq 0;\ \gamma \neq 0$.

If all the exponents in eqn. (1.2) are zero, i.e

$$\alpha = \beta = \gamma = 0 \tag{1.3}$$

then the unit of the quantity a cannot depend on the fundamental units L, T and M

$$[a] = L^0T^0M^0 = 1 \tag{1.4}$$

The quantity a which satisfies (1.3) or (1.4) is referred to as a *dimensionless quantity*.

Hence, the unit and thus the numerical value of a dimensional quantity is dependent on the choice of the fundamental units; the unit and thus the numerical value of a dimensionless quantity being independent of the choice of the fundamental units. Accordingly, dimensionless quantities maintain the same numerical values in *all* systems of fundamental units.

Consider the power product X_j formed by three dimensional quantities a_1, a_2 and a_3.

$$X_j = a_1^{x_j}a_2^{y_j}a_3^{z_j} \tag{1.5}$$

If it is possible to determine the exponents x_j y_j, and z_j so that the power product X_j becomes dimensionless, then the dimensions of a_1, a_2 and a_3 are dependent; if it is not possible the dimensions of a_1, a_2 and a_3 are independent. Clearly, the possibility of the formation of a dimensionless power product out of the quantities a_1, a_2 and a_3 having the dimensions

$$[a_1] = L^{\alpha_1}T^{\beta_1}M^{\gamma_1}$$
$$[a_2] = L^{\alpha_2}T^{\beta_2}M^{\gamma_2}$$
$$[a_3] = L^{\alpha_3}T^{\beta_3}M^{\gamma_3}$$

depends on whether the determinant

$$\Delta = \begin{vmatrix} \alpha_1 & \alpha_2 & \alpha_3 \\ \beta_1 & \beta_2 & \beta_3 \\ \gamma_1 & \gamma_2 & \gamma_3 \end{vmatrix} \tag{1.6}$$

is equal to zero or not. If $\Delta = 0$, then the formation of a dimensionless power product is possible, and thus the dimensions of a_1, a_2 and a_3 are dependent. Indeed, the power product (1.5) can be rendered dimensionless only if the system of linear and homogeneous equations

$$\alpha_1 x_j + \alpha_2 y_j + \alpha_3 z_j = 0$$
$$\beta_1 x_j + \beta_2 y_j + \beta_3 z_j = 0$$
$$\gamma_1 x_j + \gamma_2 y_j + \gamma_3 z_j = 0$$

has a solution with respect to x_j, y_j and z_j, i.e. if the coefficient determinant Δ vanishes. Conversely, if $\Delta \neq 0$, then the dimensions of a_1, a_2 and a_3 are independent. For example, if a_1, a_2 and a_3 are length, time and mass, then (according to definition) their dimensions must be independent. Indeed

$$\begin{vmatrix} 1 & 0 & 0 \\ 0 & 1 & 0 \\ 0 & 0 & 1 \end{vmatrix} \neq 0$$

If a_1, a_2 and a_3 are, say, length, velocity and acceleration, then

$$\begin{vmatrix} 1 & 1 & 1 \\ 0 & -1 & -2 \\ 0 & 0 & 0 \end{vmatrix} = 0$$

and thus the dimensions of length, velocity and acceleration are dependent. If, on the other hand, a_1, a_2 and a_3 represent length, velocity and force, then their dimensions are independent, for

$$\begin{vmatrix} 1 & 1 & 1 \\ 0 & -1 & -2 \\ 0 & 0 & 1 \end{vmatrix} \neq 0$$

and so on.

Let us now consider three quantities a_1, a_2 and a_3 which possess independent dimensions. In this case, from three equations

$$\left.\begin{aligned}[a_1] &= L^{\alpha_1}T^{\beta_1}M^{\gamma_1}\\ [a_2] &= L^{\alpha_2}T^{\beta_2}M^{\gamma_2}\\ [a_3] &= L^{\alpha_3}T^{\beta_3}M^{\gamma_3}\end{aligned}\right\} \tag{1.7}$$

the three 'unknowns' L, T and M can be solved in terms of $[a_1]$, $[a_2]$ and $[a_3]$ as follows

$$\left.\begin{aligned}L &= [a_1]^{k_1}[a_2]^{l_1}[a_3]^{m_1}\\ T &= [a_1]^{k_2}[a_2]^{l_2}[a_3]^{m_2}\\ M &= [a_1]^{k_3}[a_2]^{l_3}[a_3]^{m_3}\end{aligned}\right\} \tag{1.8}$$

Indeed, taking the logarithm of both sides of (1.7) we arrive at the following system of linear equations.

$$\left.\begin{aligned}\ln [a_1] &= \alpha_1(\ln L) + \beta_1(\ln T) + \gamma_1(\ln M)\\ \ln [a_2] &= \alpha_2(\ln L) + \beta_2(\ln T) + \gamma_2(\ln M)\\ \ln [a_3] &= \alpha_3(\ln L) + \beta_3(\ln T) + \gamma_3(\ln M)\end{aligned}\right\} \tag{1.9}$$

which indicates clearly the possibility of the solution (1.8), since the quantities a_1, a_2 and a_3 possess independent dimensions and thus the determinant

$$\Delta = \begin{vmatrix}\alpha_1 & \beta_1 & \gamma_1\\ \alpha_2 & \beta_2 & \gamma_2\\ \alpha_3 & \beta_3 & \gamma_3\end{vmatrix}$$

is different from zero.

According to (1.2) the dimension $[a]$ of *any* quantity a is given as a power product of L, T and M. On the other hand, from (1.8) it follows that the fundamental units L, T and M themselves can be expressed as power products of the units of any three quantities a_1, a_2 and a_3 having independent dimensions. But if so, the dimension of any quantity a can equally well be expressed by the units $[a_1]$, $[a_2]$ and $[a_3]$ of any three dimensionally independent quantities. In other words, the power product (1.2) is not the only way for the expression of a dimension. The dimension of a quantity a can equally well be given as

$$[a] = [a_1]^{\bar{\alpha}}[a_2]^{\bar{\beta}}[a_3]^{\bar{\gamma}} \tag{1.10}$$

where $\bar{\alpha}$, $\bar{\beta}$ and $\bar{\gamma}$ are the following linear combinations of α, β and γ

$$\left.\begin{aligned}\bar{\alpha} &= k_1\alpha + k_2\beta + k_3\gamma\\ \bar{\beta} &= l_1\alpha + l_2\beta + l_3\gamma\\ \bar{\gamma} &= m_1\alpha + m_2\beta + m_3\gamma\end{aligned}\right\} \tag{1.11}$$

Hence, any quantity a that belongs to the realm of mechanics can, in fact, be defined by means of any three independent entities a_1, a_2 and a_3. It appears that the conventional consideration of length, time and mass as the 'three pillars' of mechanics is primarily due to inborn human instinct rather than to logical reasoning. Even the number of quantities chosen as fundamental quantities does not necessarily need to be three. The number (k) of the fundamental quantities, in mechanics as well as in the other branches of physics can be selected arbitrarily[1]. However, in order to maintain a uniform method and avoid unfamiliar versions, we will always express the dimension of a quantity a in accordance with the existing convention, that is in terms of L, T and M. The dimensions of some quantities frequently used in fluid mechanics and hydraulics are given in Table 1.1

Table 1.1

Dimensional quantity	α	β	γ	Dimensional quantity	α	β	γ
Length	1	0	0	Stress (pressure)	−1	−2	1
Time	0	1	0	Density	−3	0	1
Mass	0	0	1	Specific weight	−2	−2	1
Velocity	1	−1	0	Viscosity	−1	−1	1
Acceleration	1	−2	0	Kinematic viscosity	2	−1	0
Volumetric flow rate	3	−1	0	Work (energy)	2	−2	1
Force (weight)	1	−2	1	Power	2	−3	1

1.2 Characteristic Parameters

The laws governing physical phenomena in general and mechanical phenomena in particular are expressed in the form of methematical relations among the quantities involved. Accordingly, a physical phenomenon must be defined in such a way as to be suitable for the generation of mathematical relations. A 'quantitative definition' of a physical phenomenon rests on the revealing of that set of n independent quantities

$$a_1, a_2, a_3, \ldots a_n \tag{1.12}$$

that are necessary and sufficient in order to describe that phenomenon completely.* These n independent quantities a_i which are required for a complete definition of a phenomenon are called the *characteristic parameters* (of that phenomenon); they can be dimensional or dimensionless,

* n quantities a_i are independent if none of them can be given as a function of the others. To be independent in this (conventional) sense and to possess independent dimensions (in the sense $\Delta \neq 0$) are entirely different things having no bearing on each other.

constants or variables. Usually the characteristic parameters are supposed to be dimensional quantities. However, there is no logical necessity for such a restriction (see Example 1.4).

A certain phenomenon can possess (or we can attribute to it) an unlimited number of *quantitative properties* which can be denoted by

$$A_1, A_2, A_3 \dots A_j \dots$$

or simply by A. Our awareness of a phenomenon is due to our awareness of its properties. Thus, by saying that 'the phenomenon is defined, or determined, by n characteristic parameters a_i', what we really mean is that its properties are determined by the characteristic parameters a_i. Hence, any quantitative property A of a phenomenon must be related to n characteristic parameters a_i, by a certain functional relation

$$A = f_A(a_1, a_2, \dots a_n) \tag{1.13}$$

Here the subscript A in f_A signifies that the form of the functional relation above depends on the nature of the property A under investigation. It follows that various properties of a phenomenon can be considered as various functions of the same n characteristic parameters.

When stating that the characteristic parameters a_i are necessary and sufficient for a complete definition of a phenomenon, we imply that they are necessary and sufficient in order to ensure the definition of any property A of the phenomenon that we can think of. This, however, does not mean that all n parameters a_i must necessarily appear in the expression of every property A. With respect to the definition of a single property A, n characteristic parameters a_i have to be regarded merely as sufficient. The fact that the relation (1.13) which is supposed to be valid for any A is expressed by all n parameters a_i does not contradict the preceding statement, since the form of the function f_A is not subjected to any restriction. If, from the theory or experiment, it is known that a certain property A cannot be dependent on one (or some) of the parameters a_i, then this parameter (or these parameters) should, of course, be excluded from the functional relation f_A corresponding to that property. On the other hand, if no prediction of this kind can be made, then there remains no alternative but to assume that the property A under investigation varies as a function of all n characteristic parameters a_i (until the subsequent research reveals it to be otherwise). From the relation (1.13) which appears as a function of n 'variables', one should not conclude that characteristic parameters must necessarily be variable quantities. Here we are not dealing with the quantities which vary, but with those which describe (or determine) a phenomenon. For example, when studying the terminal velocity of a particle falling in a liquid, or when determining the expression of a flow over a weir, one must remember that these phenomena are due to and thus determined by, the force of gravity. Thus, the acceleration due to gravity g must necessarily be included in the list of symbols (1.12) of these phenomena, irrespective of the fact that as far as our planet is concerned, g can be regarded as a constant quantity. Similarly, if one intends to determine a relation that will be used solely for the flow of a particular fluid (say water) taking place at a particular order of temperature (i.e. for such conditions where the density of the fluid ρ and its viscosity μ can be considered as constants) the quantities ρ and μ must still be included in the list of characteristic parameters, for the motion of the particular fluid under consideration must certainly be dependent on its density and viscosity. The reader not aquainted with the theory of dimensions might wonder why some constant parameters should be treated as 'variables' of the function f_A. The answer to this question lies in the fact that expression (1.13) is not the ultimate functional relation determining the property A, while the (dimensional) characteristic parameters are not the actual variables of a phenomenon; they are merely the 'ingredients' needed in order to form the actual variables. In the next section it will be shown that the actual variables of a physical phenomenon are dimensionless variables, formed by the characteristic parameters a_i. If the characteristic parameters a_i are selected wrongly, the dimensionless variables will automatically be wrong. If, for example, one of the characteristic parameters is omitted (e.g. because it is a constant!) then one of the dimensionless variables will be missing, and yet this dimensionless variable could indeed have been a variable quantity formed by the omitted constant parameter and some other varying parameters.

In section 1.1 it has been pointed out that it is irrelevant whether the dimension of a mechanical quantity is expressed by the units L, T and M or by the units of some other three quantities having independent dimensions. All that is relevant is the number (three) of the selected units and their independence. An entirely analogous statement can be made with respect to the characteristic parameters (and as will be seen later with respect to the dimensionless variables). It appears nature is not concerned with the meaning of the 'elements' forming her laws; all that she seems to be concerned with is the number of these 'elements' and whether they are independent. Even though the composition of the elements in the expression of a natural law varies, depending on their meaning, the mentioned properties of the laws nevertheless appear to be in harmony with the philosophical view that the basis of the physical world is not so much physical as mathematical. The fact that it is not the nature of the characteristic parameters which is relevant, but their number (n), and independence, can be demonstrated in the following manner. Consider any n independent properties $A_1, A_2, \ldots A_n$.

$$\left.\begin{aligned} A_1 &= f_1(a_1, a_2, \ldots a_n) \\ A_2 &= f_2(a_1, a_2, \ldots a_n) \\ &\cdot \\ &\cdot \\ &\cdot \\ A_n &= f_n(a_1, a_2, \ldots a_n) \end{aligned}\right\} \quad (1.14)$$

From n independent eqns. above n quantities a_i can be solved (in principle) as follows

$$\left.\begin{aligned} a_1 &= \bar{f}_1(A_1, A_2, \ldots A_n) \\ a_2 &= \bar{f}_2(A_1, A_2, \ldots A_n) \\ &\cdot \\ &\cdot \\ &\cdot \\ a_n &= \bar{f}_n(A_1, A_2, \ldots A_n) \end{aligned}\right\} \quad (1.15)$$

Consider now any property A which is not one of n selected properties $A_1, A_2, \ldots A_n$. We have

$$A = f_A(a_1, a_2, \ldots a_n) \quad (1.16)$$

Substituting the values of $a_1, a_2, \ldots a_n$ given by (1.15) into (1.16) we obtain

$$A = f_A(\bar{f}_1, \bar{f}_2, \ldots \bar{f}_n) \quad (1.17)$$

Here each $\bar{f}_i$ consists only of $A_1, A_2, \ldots A_n$, and thus the right hand side of (1.17) is, in fact, a function only of $A_1, A_2, \ldots A_n$

$$A = \bar{f}_A(A_1, A_2, \ldots A_n) \quad (1.18)$$

The relation (1.18) implies that any property A which is given (by (1.16)) in terms of n parameters $a_1, a_2, \ldots a_n$, can equally well be given in terms of any n independent properties $A_1, A_2, \ldots A_n$. But this is another way of saying that any n independent quantities related to a phenomenon can be selected as its characteristic parameters.

1.3 Dimensionless Expression of a Natural Law

A functional relation such as (1.13) i.e.

$$A = f_A(a_1, a_2, \ldots a_n)$$

is supposed to represent one of the laws governing a natural phenomenon, i.e. one of the laws of the physical world which is supposed to exist independently of the human mind. Accordingly, the numerical values of a correctly expressed natural law cannot depend on any functioning (activity) of the human mind. On the other hand, if the property A in the expression above is a dimensional quantity, then its numerical value (that is the numerical value of the function f_A) will obviously depend on our arbitrary choice of fundamental units. Hence, a dimensional expression such as the functional relation above cannot be a correct way of expressing a natural law. From the content of Section 1.1 it is clear that only dimensionless quantities possess numerical values that are independent of arbitrary choice of fundamental units. Thus, the correct expression of a natural law can only be dimensionless. One can arrive at the same conclusion from the principles of the theory of relativity, which treats the length, time and mass as relative entities which do not possess an absolute measure. But if so, then the expression of a natural law, which is supposed to be of an absolute (universal) validity, cannot be identified with one of these three relative entities or with a (dimensional) combination thereof.

The dimensionless equivalent of the dimensional relation such as that shown above is given by the procedure which is usually referred to as the 'π-theorem' and which can be described as follows[1,2,3]. Among n characteristic parameters

$$a_1, a_2, \ldots a_n$$

any three parameters are selected which possess independent dimensions (see Section 1.1). These selected three parameters will be referred to henceforward as *basic quantities*. We can assume, for the sake of simplicity, that the basic quantities of the above set of characteristic parameters are just the first three parameters a_1, a_2 and a_3. These three basic quantities are combined with the remaining $n - 3$ parameters in the following $N = n - 3$ independent power products

$$\left.\begin{aligned} X_1 &= a_1^{x_1} a_2^{y_1} a_3^{z_1} a_4^{m_1} \\ X_2 &= a_1^{x_2} a_2^{y_2} a_3^{z_2} a_5^{m_2} \\ &\cdot \\ &\cdot \\ &\cdot \\ X_N &= a_1^{x_N} a_2^{y_N} a_3^{z_N} a_n^{m_N} \end{aligned}\right\} \quad \text{(where } N = n - 3\text{)} \qquad (1.19)$$

* $N = n - 3$ power products $X_1, X_2, \ldots X_N$ are independent, that is none of them can be expressed in terms of the remaining $N - 1$ products, because each of them contains one, independent, characteristic parameter ($a_4, a_5, \ldots a_n$) that does not appear in any other of the remaining $N - 1$ power products.

which are the *dimensionless variables* of the phenomenon given by n characteristic parameters $a_1, a_2, \ldots a_n$. In the expression above, the exponents $m_1, m_2, \ldots m_N$ can be selected arbitrarily. Depending on the selected values of $m_1, m_2, \ldots m_N$ the exponents x_j, y_j and z_j $(j = 1, 2, \ldots N)$ must be determined so that each of the power products X_j becomes dimensionless. In order to determine three unknown exponents x_j, y_j and z_j of the power product X_j, we have the following system of three eqns.

$$\begin{aligned} \alpha_1 x_j + \alpha_2 y_j + \alpha_3 z_j &= -\alpha_{j+3} m_j \\ \beta_1 x_j + \beta_2 y_j + \beta_3 z_j &= -\beta_{j+3} m_j \\ \gamma_1 x_j + \gamma_2 y_j + \gamma_3 z_j &= -\gamma_{j+3} m_j \end{aligned}$$

where α_i, β_i and γ_i are known from the dimension formula

$$[a_i] = L^{\alpha_i} T^{\beta_i} M^{\gamma_i}$$

of the parameter a_i. The system of three linear eqns. for x_j, y_j and z_j can certainly be solved, for its coefficient determinant Δ is different from zero (by virtue of the fact that the basic quantities a_1, a_2 and a_3 have independent dimensions).

The dimensionless expression of the property A under investigation is given in a similar manner by the power product

$$\Pi_A = a_1^{x_A} a_2^{y_A} a_3^{z_A} A \tag{1.20}$$

Here also the exponents x_A, y_A and z_A must be determined (depending on the dimension of the property A) so that the power product Π_A becomes dimensionless. Now the dimensionless combination Π_A corresponding to the property A is a certain function of $N = n - 3$ dimensionless variables $X_j (= 1, 2, \ldots N)$, i.e.

$$\Pi_A = \varphi_A(X_1, X_2, \ldots X_N) \tag{1.21}$$

which is the dimensionless equivalent of the functional relation (1.13).

The subscript A in φ_A signifies that the form of the function φ_A depends on the nature of the property A under investigation. The dimensionless relationship (1.21) has, by comparison with its dimensional counterpart (1.13), the following advantages:

(i) it is a correct version of a natural law,
(ii) it is a function of a reduced (by three) number of variables, which implies that the form of φ_A can be determined with fewer difficulties than that of f_A,
(iii) its numerical value does not depend on the system of units,
(iv) and, as will be seen in the next chapter, its variables X_j are at the same time the *criteria of similarity*, of the phenomenon given by the characteristic parameters a_i.

Observe (from (1.19)) that the parameter a_4 appears only in the expression of the variable X_1, the parameter a_5 only in X_2 and so on. It is said, therefore, that 'the influence of the parameter a_4 is taken into account by the variable X_1', 'the influence of a_5 by X_2', and so on: the influence of the parameters a_1, a_2 and a_3 (which are selected as basic quantities) not being taken into account by any of the variables X_j. If the consideration of the influence of a parameter, a_i, by a special variable is desired, then such a parameter must not be selected as a basic quantity.

In the case of dimensionless variables it is again the number $N = n - 3$ and the independence of the dimensionless variables X_j that matters and not their nature. This fact can easily be proved in a manner completely analogous to the demonstration leading to eqn. (1.18) (replace in (1.14) A_k by Π_{A_k}, a_i by X_j and f_k by φ_k and follow the same pattern of derivation). The fact that any three parameters possessing independent dimensions can be selected as basic quantities, and thus that the dimensionless variables X_j can be formed in various ways is an advantage that can be used to serve some practical purposes. For example, if the form of the function φ_A is to be determined by experimental measurements, then it is always desirable that when one of the dimensionless variables varies, the others remain constant. This can be achieved simply by not selecting the parameters which will be varied during the experiment as basic quantities (for the basic quantities appear in the expression of every dimensionless variable, whereas each of the remaining parameters in only one).

It should be pointed out that if the phenomenon involves some irregular shapes, one does not usually attempt to describe this irregular geometry by special parameters. Indeed, the description of the shape of an arbitrary curve or of a surface cannot be given adequately by a limited number of parameters. Consider, for example, the flow cross-section shown in Fig. 1.1a and let us attempt to describe the shape of this irregular cross-section by introducing some special geometric parameters. We might think that if we give the flow width B, the maximum depth h, the angles α_1, and α_2, the distances l_1 and l_2 (corresponding to, say, the middle of the depth h), and if we specify the position of the lowest point of the cross-section by introducing the distance l_3, then the resulting set of seven 'shape parameters'

$$B,\ h,\ \alpha_1,\ \alpha_2,\ l_1,\ l_2,\ l_3$$

will at least approximately describe the shape of the flow cross-section. To show that such an assumption is too optimistic, it is sufficient to observe that the cross-section in Fig. 1.1b (which corresponds to precisely the same values of the above seven parameters) possesses a shape that cannot be regarded even as 'approximately equal' to that of Fig. 1.1a. The properties of the flow in Figs. 1.1a and 1.1b are completely different; this is entirely due to the differing geometric shapes of the cross-sections. Hence, the introduction of as many as seven special 'shape parameters', did not, in fact, serve the purpose.

On the other hand, if the shape parameters are omitted, it implies that a particular form of the functions f_A and φ_A can correspond only to a particular shape (or shapes) involved in the phenomenon under consideration. For example, if the form of a function φ_A is determined from the flow

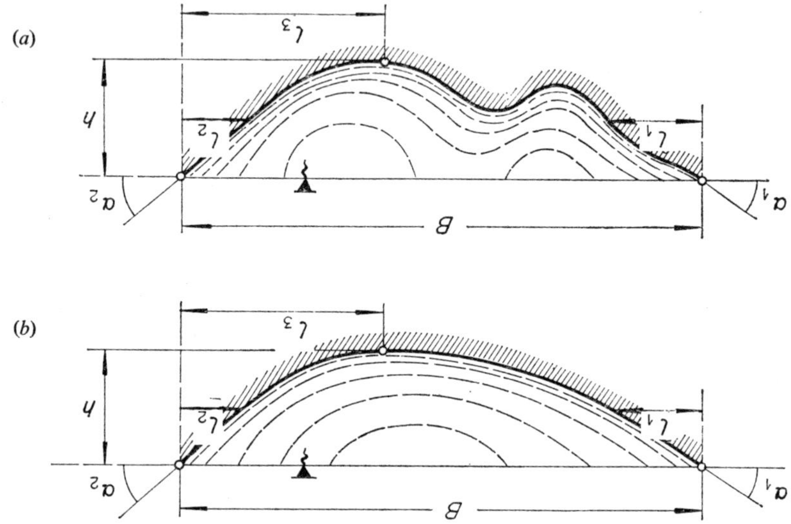

FIG. 1.1

measurements carried out in a pipe having a circular cross-section and sand roughness, then we must not expect that the function φ_A will maintain its form for the case of a 'quadratic' or 'triangular' pipe cross-section and for the roughness having the texture that is not geometrically similar to that of tightly packed sand grains. In many cases, it is not possible to describe the geometry by words such as circular, quadratic, parabolic, etc. In such cases we simply produce drawings of the shapes which are involved in the phenomenon and for which a function φ_A is supposed to be valid. For example, we say that the particular form

$$\Pi_A = \bar{\varphi}_A(X_1, X_2, \ldots X_n)$$

is valid for the cross-section geometry shown in Fig. 1.1a; whereas the form

$$\Pi_A = \bar{\bar{\varphi}}_A(X_1, X_2, \ldots X_n)$$

is valid for that in Fig. 1.1b.

It frequently happens that an idea can best be introduced by illustration in specific examples. In the present book examples are used not only in order to show the application of the theory already introduced, but also as a suitable medium for the introduction of new ideas. The reader should therefore consider the examples as part of the text.

EXAMPLE 1.1 Flow over a weir

Consider the flow over a weir of a specified geometry. For example, the

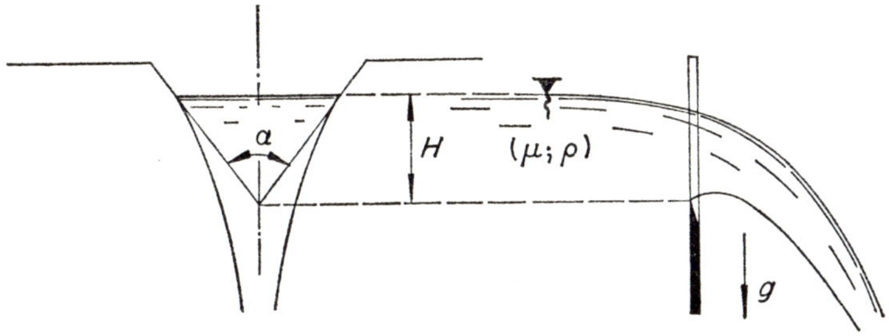

FIG. 1.2

weir might be of a symmetrical triangular shape with a given value of the angle α (Fig. 1.2). This weir flow is completely determined

(a) by the nature of the fluid
(b) by the 'weir load' H, and
(c) by the acceleration due to gravity g (which brings the flow into existence).

As far as the study of mechanics is concerned, the nature of the fluid is completely determined by the numerical values of its viscosity μ and the density ρ. If the values of μ and ρ are given, then with regard to the mechanical considerations, the fluid is given. Hence, the set of characteristic parameters of the weir flow under consideration can be formed by the following four independent quantities

$$\mu, \rho, H, g \tag{1.22}$$

Any mechanical quantity A related to this flow phenomenon must be a certain function of $n = 4$ characteristic parameters above, i.e.

$$A = f_A(\mu, \rho, H, g) \tag{1.23}$$

Let us assume that the property A is the flow rate (or discharge) Q. According to (1.23) the value of Q must be given by

$$Q = f_Q(\mu, \rho, H, g) \tag{1.24}$$

The dimensionless version Π_Q of the property Q must be a function of $N = 4 - 3 = 1$ (one) dimensionless variable X_1, i.e.

$$\Pi_Q = \varphi_Q(X_1) \tag{1.25}$$

The parameters ρ, H and g have independent dimensions (see Table 1.1) and therefore they can be selected as basic quantities. By doing so, and considering (1.19) and (1.20) we obtain

$$X_1 = \rho^1 H^{\frac{3}{2}} g^{\frac{1}{2}} \mu^{-1} = \frac{\rho H^{\frac{3}{2}} g^{\frac{1}{2}}}{\mu} \tag{1.26}$$

and

$$\Pi_Q = \rho^0 H^{-\frac{5}{2}} g^{-\frac{1}{2}} Q = \frac{Q}{H^{\frac{5}{2}} g^{\frac{1}{2}}}$$

Thus

$$\frac{Q}{H^{\frac{5}{2}} g^{\frac{1}{2}}} = \varphi_Q \left(\frac{\rho H^{\frac{3}{2}} g^{\frac{1}{2}}}{\mu} \right) \tag{1.27}$$

or

$$Q = m\sqrt{(2g)}H^{\frac{5}{2}} \tag{1.28}$$

which is a well-known formula of hydraulics, where the dimensionless weir coefficient m is a certain function of the combination X_1 which reflects the influence of viscosity

$$m = \bar{\varphi}_Q \left(\frac{\rho H^{\frac{3}{2}} g^{\frac{1}{2}}}{\mu} \right) \qquad \left(\text{with } \bar{\varphi}_Q = \frac{1}{\sqrt{2}} \varphi_Q \right) \tag{1.29}$$

The form of the function $\bar{\varphi}_Q$ of the dimensionless variable X_1 is dependent on the geometry of the weir and can be determined only by experiment. Observe that the variation of

$$X_1 = \frac{\rho H^{\frac{3}{2}} g^{\frac{1}{2}}}{\mu}$$

can be achieved by varying *only* H, that is by using always the same fluid (μ = const, ρ = const) and the same weir. If for various values of H (and thus X_1) the corresponding values of Q are measured and the dimensionless ratio $Q/\sqrt{(2g)}H^{5/2}$, meaning m, is computed, then by plotting the values of m versus the corresponding values of the combination X_1 we arrive at the sequence of experimental points forming the experimental curve of the relation $m = \bar{\varphi}_Q(X_1)$ (see eqn. (1.29)). Even if the experimental measurements were carried out by using the same fluid (say water) and the same weir, the experimental curve $m = \bar{\varphi}_Q(X_1)$ determined would be valid for all fluids and for all geometrically similar weirs.

The dimensionless variable X_1 which can be written as

$$X_1 = \frac{\rho H \sqrt{(gH)}}{\mu} \tag{1.30}$$

can be regarded as a Reynolds number, as the multiplier $\sqrt{(gH)}$, which has the dimension of the velocity, can be regarded as a *typical velocity* of the weir flow. Accordingly, the dimensionless variable X_1 can be interpreted as the ratio of the inertia force (which is proportional to the specific mass ρ) to the viscous friction force (which is proportional to the viscosity μ).

$$X_1 = \frac{\text{inertia force}}{\text{viscous force}} \tag{1.31}$$

This relation indicates that the larger the value of the Reynolds number X_1, the smaller the magnitude of the viscous force in comparison to the inertia force, and thus the smaller the influence of the Reynolds number X_1 on the progress of the phenomenon, for it is the Reynolds number X_1 which takes into account the role of μ (and which is present because the parameter μ is present). Observe, from (1.30) that for a given fluid ($\mu = \text{const}$, $\rho = \text{const}$) the value of X_1 increases when H increases. Hence, for sufficiently large weir loads H, the Reynolds number X_1 must cease to be the variable of the phenomenon, which implies that the function (1.29) must degenerate into $m = \text{const}$, while eqn. (1.28) must become

$$Q = \text{const}\ \sqrt{(2g)}H^{\frac{5}{2}} \tag{1.32}$$

Here, the value of the constant depends only on the geometry of the weir. For example, if the symmetrical triangular weir has $\alpha = \pi/2$ (90° *V*-notch or 'Thomson weir') then, as is known from the courses on hydraulics, the constant in the equation above possesses the following, experimentally determined, value

$$\text{const} = 0{\cdot}316 \tag{1.33}$$

Observe that if μ is excluded from the set (1.22), then $n = 3$, and thus $n - 3 = 0$, which implies that Π_Q is a function of no variables or that

$$\Pi_Q = \frac{Q}{H^{\frac{5}{2}} g^{\frac{1}{2}}} = \overline{\text{const}} \qquad (\text{with } \overline{\text{const}} = \sqrt{2}\text{const})$$

which in principle is nothing else but the relation (1.32). Hence, the relation (1.32) can also be obtained by the formal procedure.

Let us now assume that the weir has a rectangular cross-section (Fig. 1.3). In this case the weir width b is an additional characteristic parameter and we arrive at the following set

$$\mu, \rho, H, g, b \tag{1.34}$$

Thus

$$\Pi_Q = \varphi_Q(X_1, X_2)$$

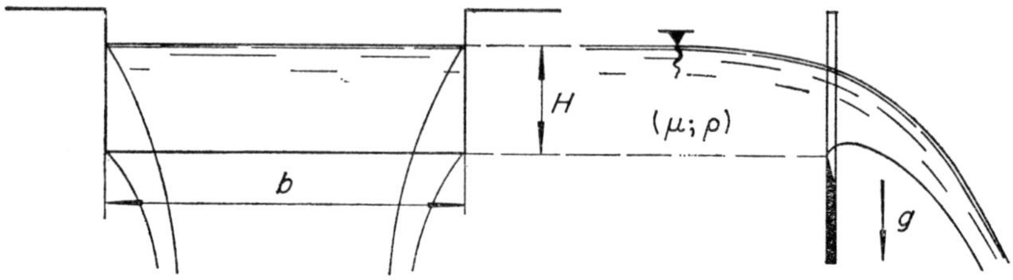

FIG. 1.3

If ρ, H and g are selected again as basic quantities, then X_1 is given again by the expression (1.26), whereas X_2 being given by

$$X_2 = \rho^0 H^{-1} g^0 b = \frac{b}{H} \tag{1.35}$$

Accordingly,

$$\frac{Q}{H^{\frac{5}{2}} g^{\frac{1}{2}}} = \varphi_Q \left(\frac{\rho H^{\frac{3}{2}} g^{\frac{1}{2}}}{\mu}; \frac{b}{H} \right) \tag{1.36}$$

In engineering practice the order of X_1 is usually large, and therefore the relation (1.36) usually appears in the form

$$\frac{Q}{H^{\frac{5}{2}} g^{\frac{1}{2}}} = \varphi_Q \left(\frac{b}{H} \right) \tag{1.37}$$

If there is no side contraction, then for the given H the flow rate Q increases proportionally with b. But this implies that φ_Q must be a linear function of its variable X_2, i.e. that

$$\varphi_Q \left(\frac{b}{H} \right) = \overline{\text{const}} \, \frac{b}{H} \tag{1.38}$$

must be valid. Substituting (1.38) in (1.37) we arrive at the following expression

$$Q = \text{const} \, . \, b\sqrt{(2g)} \, . \, H^{\frac{3}{2}} \tag{1.39}$$

$$(\text{with const} = \overline{\text{const}}/\sqrt{2})$$

which is also a very well-known weir flow formula.

In the considerations above it was assumed that the weir height p is large in comparison to the weir load H, i.e. that $H/p \to 0$.

However, these considerations can easily be generalized to the case of finite values of H/p, by introducing p as a characteristic parameter, and thus by introducing $X_3 = H/p$ as an additional dimensionless variable.

EXAMPLE 1.2 Stationary and uniform flow in a closed conduit, Flexibility in the interpretation

Let us now consider the stationary and uniform (parallel) flow of a real fluid in a rough closed conduit. It is assumed that the shape of the conduit cross-section is specified and that the geometry (texture) of its roughness is given. Such a flow phenomenon is completely determined:

(a) by the nature of the fluid (i.e. μ and ρ),
(b) by the absolute size of the system (which can be given by any linear characteristic of the flow cross-section, and thus by the length $R = \dfrac{\omega}{\chi}$ (called hydraulic radius),
(c) by the absolute size k of the protrusions forming the roughness, and
(d) by the kinematic state of motion (which can be given by the average velocity v of the flow).

Thus the set of characteristic parameters can be formed by the following five quantities

$$\mu, \rho, R, k, v \tag{1.40}$$

Any mechanical quantity A related to the flow under consideration must be expressible as a function of these five characteristic parameters. Let us, for example, consider as A the average shear stress τ_o acting on the flow boundary. We have

$$\tau_o = f_{\tau_0}(\mu, \rho, R, k, v) \tag{1.41}$$

The parameters ρ, R and v possess independent dimensions and thus they can be selected as basic quantities. By doing so, and considering (1.19) we determine the following $n - 3 = 5 - 3 = 2$ dimensionless variables

$$\left.\begin{aligned} X_1 &= \rho^1 R^1 v^1 \mu^{-1} = \frac{vR}{\nu} \\ X_2 &= \rho^0 R^{-1} v^0 k^1 = \frac{k}{R} \end{aligned}\right\} \tag{1.42}$$

$$\left(\text{where } \nu = \frac{\mu}{\rho} \text{ kinematic viscosity}\right)$$

The first variable is the Reynolds number, the second being relative roughness. Considering (1.20) we determine the following dimensionless version of the average shear stress τ_o

$$\Pi_{\tau_0} = \rho^{-1}R^0v^{-2}\tau_o = \frac{\tau_o}{\rho v^2} \tag{1.43}$$

Hence, the dimensionless equivalent of (1.41) is

$$\Pi_{\tau_0} = \varphi_{\tau_0}(X_1; X_2)$$

i.e.

$$\frac{\tau_o}{\rho v^2} = \varphi_{\tau_0}\left(\frac{vR}{\nu}; \frac{k}{R}\right) \tag{1.44}$$

The relation (1.44) can be expressed in the form of a friction formula

$$\tau_o = \frac{1}{c^2}\rho v^2 \tag{1.45}$$

where

$$\frac{1}{c^2} = \varphi_{\tau_0}\left(\frac{vR}{\nu}; \frac{k}{R}\right) \tag{1.46}$$

Hence, without making any use of knowledge on fluid mechanics we have derived the friction formula, and what is more, we have revealed that the dimensionless friction coefficient c must be a certain function of the Reynolds number and the relative roughness. The reader acquainted with hydraulics will know that this information, which is explained by the theory of dimensions in one page of simple operations, can only be derived by the traditional methods of hydraulics, in a comprehensible and convincing manner, after an extensive preparatory study.

The square root of the ratio τ_o/ρ which has the dimension of a velocity (see Table 1.1) is referred to as shear velocity and is denoted as follows:

$$v_* = \sqrt{\left(\frac{\tau_o}{\rho}\right)} \tag{1.47}$$

Shear velocity v_* is thus not an actual velocity which can be measured by the pilot tube somewhere in the flow; it is merely a convenient notation which gives the possibility of treating the shear stress τ_o in a velocity like way.

Using this notation the friction formula (1.45) can be brought into the form

$$c = \frac{v}{v_*} \tag{1.48}$$

In the case of a circular pipe and the roughness formed by equal size sand grains tightly glued on the inner surface of the pipe (sand roughness), the variation of

$$\lambda = \frac{8}{c^2} \tag{1.49}$$

with the Reynolds number, and relative roughness is given by the family of experimental curves of J. Nikuradse (Fig. 1.4).

The coefficient λ comes from the Darcy-Weisbach equation

$$J = \lambda \frac{1}{D} \frac{v^2}{2g}$$

where J is the energy gradient while D is the diameter of the circular pipe. Considering that for circular pipes $\tau_0 = \rho v_*^2 = \gamma JD/4$, eliminating J, and substituting $v/v_* = c$ we arrive at the relation (1.49) between c and λ.

For the case of a circular cross-section, we have

$$\frac{vR}{\nu} = \frac{1}{4}\frac{vD}{\nu} \quad \text{and} \quad \frac{k_s}{R} = 4\,\frac{k_s}{D} \tag{1.50}$$

which explains why the values of λ are plotted, in Fig. 1.4, simply as functions of vD/ν and k_s/D. Here the symbol k_s stands for the height of sand roughness; this height is equal to the diameter of equal size sand grains tightly glued on the flow boundary. In future, the height of any roughness having the geometry (texture) statistically identical to that of the sand roughness will invariably be denoted by k_s.

The set of characteristic parameters (1.40) can be interpreted in a variety of ways. In fact, this set can represent any flow phenomenon which depends on the nature of the fluid on two lengths, and on the kinematical state of the fluid motion. Let us write the set (1.40) in the following manner (more suitable for different interpretations)

$$\mu, \rho, l_1, l_2, U \tag{1.51}$$

The above set of parameters can be interpreted, for example, as the set representing the flow in a smooth straight conduit having a rectangular cross-section (Fig. 1.5a). In this case l_1 and l_2 are the dimensions of the rectangular cross-section, whereas U can be regarded as the average velocity of the flow. Similarly, the set (1.51) can represent the flow in a smooth pipe-bend having, say, a circular cross-section of the diameter l_1 and the curvature radius l_2 (Fig. 1.5b). Figs. 1.5c and 1.5d show flows past obstacles possessing certain specified shapes. The obstacle in the former figure is smooth, but it is in the vicinity of a smooth boundary (the length l_2 being the distance from the boundary); the obstacle of the latter figure

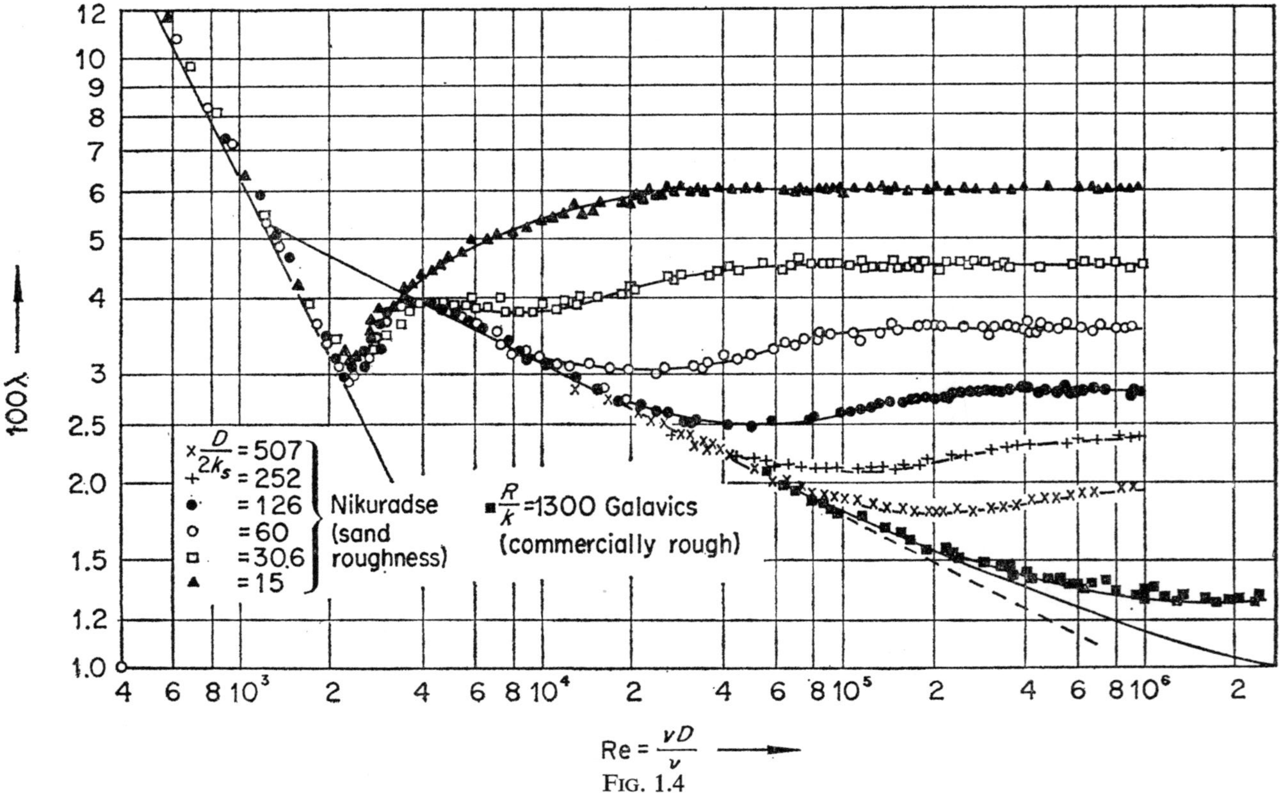

FIG. 1.4

(After H. Schlichting, *Boundary Layer Theory*, McGraw-Hill Book Co. Inc., 6th ed., 1968)

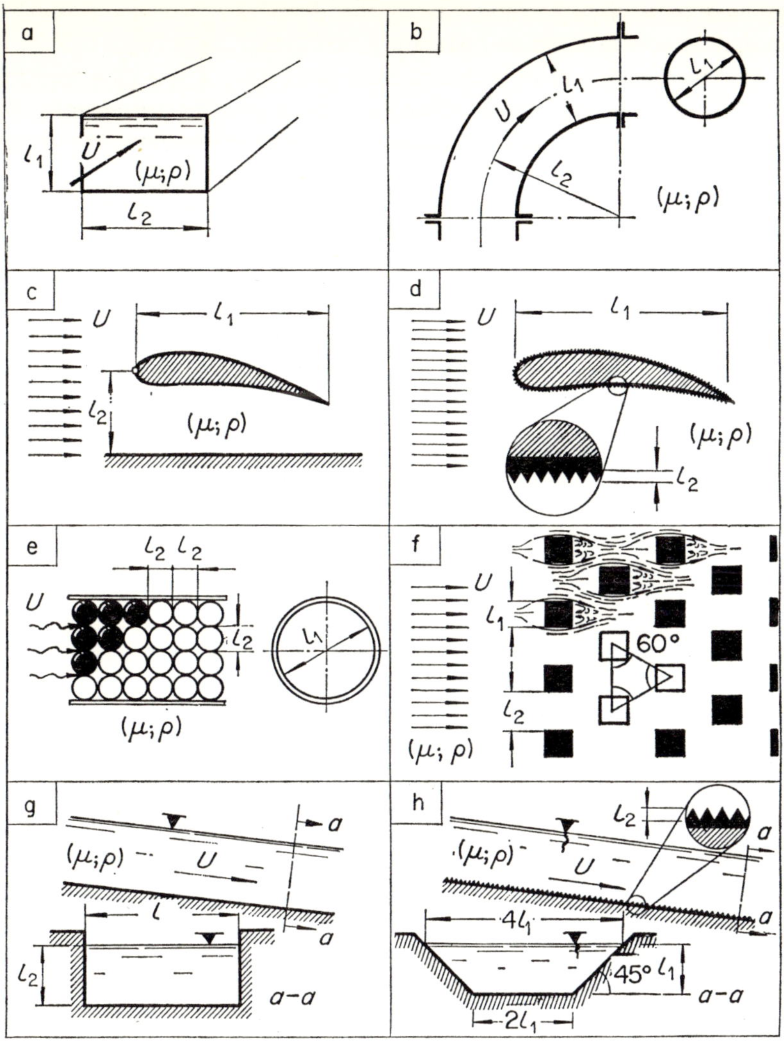

FIG. 1.5

is within an infinite fluid, but it is rough (the length l_2 being the size of its roughness). Clearly, both of these flows can be completely described by the kinematical state of the approaching flow, by the nature of the fluid and by the lengths l_1 and l_2, and thus they can both be represented by the set of parameters (1.51). Fig. 1.5e illustrates the flow through a porous material of a given geometry (shape and configuration of the grains) contained in a circular tube having the diameter l_1, which cannot be regarded as very large in comparison to the grain size l_2. Fig. 1.5f shows the flow past a system of identical bars possessing a specified shape of their cross-section and a well defined geometry of their arrangement but an adjustable (variable) distance l_2. It is obvious that the flow phenomena implied by Figs. 1.5e and 1.5f can also be represented by the set (1.51). Consider, finally, the uniform flow in a smooth open channel having, say, a rectangular cross-section (Fig. 1.5g) or the uniform flow in a rough open channel possessing a certain specified geometry of its cross-section (Fig. 1.5h). Both of these channel flows are determined by two lengths l_1 and l_2, by the nature of the fluid, and by the kinematical state of the fluid motion. Thus, they can also be described by the set of characteristic parameters (1.51), and by the dimensionless variables

$$\frac{Ul_1}{\nu} \quad \text{and} \quad \frac{l_1}{l_2} \tag{1.52}$$

$$\left(\text{with } \nu = \frac{\mu}{\rho}\right)$$

or by any two independent combinations thereof.

No limit can be drawn with regard to the physical picture that can be represented by a set of characteristic parameters of a particular nature, and thus by a particular set of dimensionless variables. In Section 1.2 we have seen that there are an unlimited number of properties $A_1, A_2, A_3, \ldots$, of a phenomenon, which are given by the same set of n characteristic parameters a_i, and thus by the same set of $N = n - 3$ dimensionless variables X_j defining a particular phenomenon. From the examples above it follows that there is an unlimited variety of the phenomena which can be represented by the same set of characteristic parameters and the same set of dimensionless variables.

EXAMPLE 1.3 Stationary and uniform flow in an open channel

In the previous Example it has been pointed out that the uniform flow in an open channel (Figs. 1.5g and 1.5h) can be represented by the dimensionless variables (1.52), and thus by the relation (1.48) which can be written as

$$v = c\sqrt{\left(\frac{\tau_0}{\rho}\right)}. \tag{1.53}$$

In the case of uniform flow in an open channel, the value of the average shear stress τ_0 acting on the flow boundary can be determined from the static equilibrium of the fluid element ABCD shown in Fig. 1.6. Here the vector $\vec{g}$ (acceleration due to gravity) can be considered by its two components

$$g_x \approx gS \quad \text{and} \quad g_y \approx g^* \tag{1.54}$$

where S is the slope of the uniform flow. The component g_x (acting per unit mass of the fluid in the direction x of the flow) produces the following magnitude of the tractive force, F_x, acting on the element ABCD

$$F_x = g_x \rho \omega \, \delta x = \gamma S \omega \, \delta x \tag{1.55}$$

where ω is the area of the cross-section, while $\gamma = \rho g$ is the specific weight of the fluid. The component g_y generates the hydrostatic pressure which

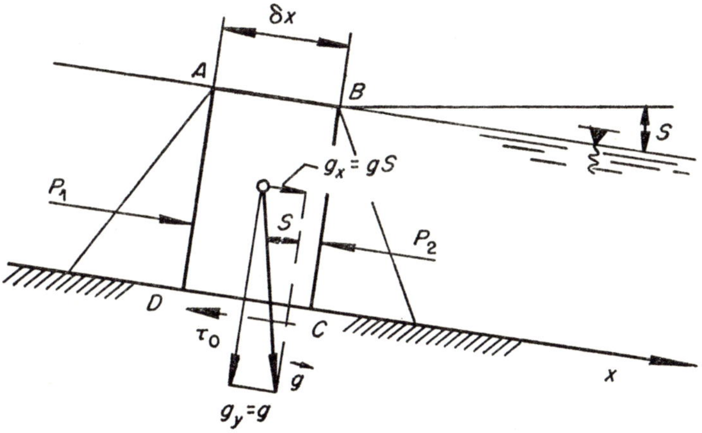

FIG. 1.6

has no bearing on the progress of the uniform flow under consideration: indeed the difference of the hydrostatic pressure forces $P_1 - P_2 = 0$. Hence, the tractive force F_x is brought into equilibrium by the average sheer stress τ_o acting on the area $\chi \, \delta x$ (where χ is the wetted perimeter), i.e.

$$\gamma S \omega \, \delta x = \tau_o \chi \, \delta x \tag{1.56}$$

is valid, which gives at once the well-known relation

$$\tau_o = \gamma S R$$

$$\text{(with } R = \omega/\chi \text{ hydraulic radius)} \tag{1.57}$$

Substituting this value of τ_o in (1.53) we obtain the following expression for the average velocity of uniform flow in an open channel

$$v = c\sqrt{(gSR)} \tag{1.58}$$

* The practical orders of the slope S are so (sufficiently) small that ($\approx$) can be regarded simply as ($=$) for all practical purposes. Accordingly, henceforth ($=$) will be used.

This expression can be written in the dimensionless form as follows

$$\frac{v^2}{g_x R} = c \tag{1.59}$$

Hence, in the case of a uniform flow, the dimensionless friction factor

$$c = \left[\varphi_{\tau_0}\left(\frac{vR}{\nu}; \frac{k}{R}\right)\right]^{-\frac{1}{2}} = \varphi_c\left(\frac{vR}{\nu}; \frac{k}{R}\right)$$

is equal to the Froude number formed by the component g_x of the acceleration due to gravity g.

It follows that the well-known Chézy formula

$$v = C\sqrt{(RS)} \tag{1.60}$$

which contains a dimensional 'Chézy coefficient' C is, with regard to dimensional considerations, an imperfect version of (1.58). From comparison between (1.58) and (1.60) we arrive at the following relation between C and c

$$C = c\sqrt{g} \tag{1.61}$$

Since the value of the dimensional multiplier $\sqrt{g}$ varies from one system of units to another, the value of the Chézy coefficient C also varies, depending on the units. This causes inconvenience when using the tables in practice. The utilisation of the relation (1.58), given in terms of the dimensionless friction factor c, is not handicapped by this practical inconvenience. From the theoretical point of view the relation (1.58) is also preferable because it indicates *how* the velocity v of the flow depends on the generator of the flow $g_x = gS$. Dividing the values of C, given by a table corresponding to a particular system of units, by the value $\sqrt{g}$ in the same system of units, we can obtain the table of c values that will be valid in any system of units.

One should not infer that the relation (1.58) can be obtained only from the equilibrium analysis of the system. One can arrive at (1.58) from purely dimensional considerations. Indeed, the uniform channel flow under consideration can be defined by the characteristic parameters (1.51), or more specifically by

$$\mu, \rho, R, k, v \tag{1.62}$$

and thus by the dimensionless variables

$$\frac{vR}{\nu}; \frac{k}{R} \tag{1.63}$$

Hence, the dimensionless versions of the quantity $g_x = gS$ can be expressed as follows

$$\Pi_{g_x} = \rho^0 R^1 v^{-2} g_x = \frac{Rg_x}{v^2} = \frac{gSR}{v^2} = \varphi_{g_x}\left(\frac{vR}{\nu}; \frac{k}{R}\right) \tag{1.64}$$

which is nothing else but (1.58), if the function Π_{g_x} is denoted by $1/c^2$.

In the considerations leading to (1.64) we have chosen v as an independent variable (characteristic parameter). Accordingly g_x, τ_0, v_*, Q . . . etc. must be regarded as the functions (properties) of the phenomenon. One could, of course, choose say, g_x, rather than v, as an independent variable. In this case v would appear as a function of the phenomenon. Such substitutions cannot make any difference in principle, for *any n* independent quantities can be selected as characteristic parameters.

EXAMPLE 1.4 Non-stationary and non-uniform flow in a non-prismatic channel

Let us now consider non-uniform flow in an open channel. We assume that the geometry of the channel (which need not necessarily be prismatic), and that the texture of its roughness, are specified. Treating this non-uniform flow, first as stationary, we can describe it:

(a) by the nature of the fluid (i.e. by μ and ρ),
(b) by the absolute size of the flow (which can be given, for example, by the hydraulic radius R_o corresponding to a section O – O that can be chosen as typical),
(c) by the size k of the channel roughness,
(d) by the kinematic state of motion (which can be given, for example, by the flow rate Q),
(e) by the slope S of the channel bed, and
(f) by the absolute value of the acceleration due to gravity $g_y \approx g$.

Thus, the set of characteristic parameters can be formed by the following seven quantities

$$\mu, \rho, R_o, k, Q, S, g \tag{1.65}$$

In case the reader might doubt whether one should still introduce the flow rate Q as a characteristic parameter if the nature of the fluid, slope, size, roughness, and geometry of the channel are already specified, one can show, as an example, the non-uniform flow in Fig. 1.7a. Here the position of the gate, and thus the value of R_0 at the typical section O = O is fixed, so is the level (elevation) E_1 of the tailwater. However, the level E_o can possess any value. Thus the velocity v_o at the section O = O, and consequently the flow rate Q can vary independently. It is clear that this independently varying value of Q (or of v_o) will affect what happens to the non-uniform flow between the sections $O = O$ and $I = I$, and thus

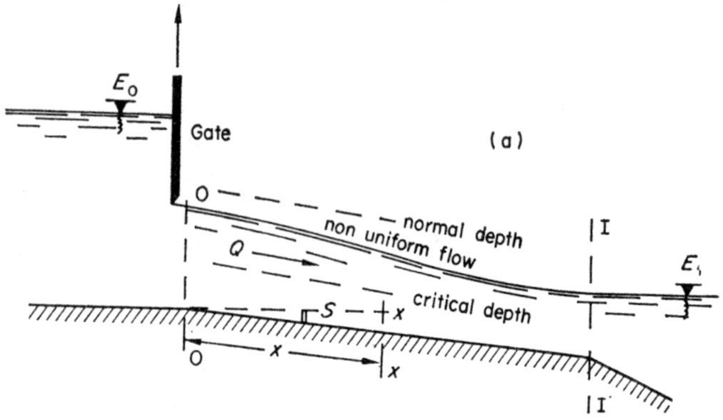

FIG. 1.7a

it has to be included in the set of characteristic parameters. The introduction of Q becomes redundant—or in more general terms, the number of the characteristic parameters (1.69) can be reduced by one—if the non-uniform flow approaches asymptotically to a uniform flow in one of its upstream or downstream ends (Fig. 1.7b). In this case, the typical section $O = O$ can be identified with any of the sections of the uniform part of

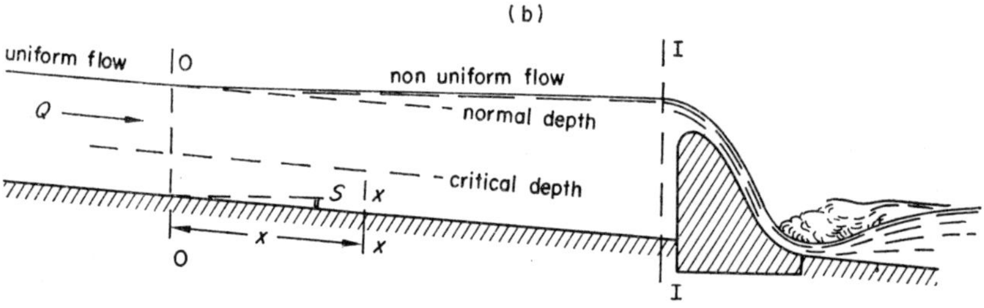

FIG. 1.7b

the flow where Q obviously is a certain function of μ, ρ, R_o, k and S, and thus it is no longer independent. However, in the following we will deal with the general case of non-uniform flow defined by $n = 7$ characteristic parameters.

Because it is not the nature of the characteristic parameters but their number ($n = 7$) and independence that is relevant, one can modify the set (1.69) so as to make it more suitable for comparison with the set (1.62), which represents the special case of the non-uniform flow phenomenon under investigation. Since the geometry of the cross-section is specified, the consideration of R_o and Q is equivalent to that of R_o and $Q/R_o^2 \sim Q/\omega_o = v_o$ (where ω_o and v_o represent the area and the average velocity of the section O — O). Replacing Q by v_o we arrive at

$$\mu, \rho, R_o, k, v_o, S, g \tag{1.66}$$

which differs from (1.62) because of the two additional parameters S and g. Selecting ρ, R_o and v_o as basic quantities, we obtain the following $n - 3 = 4$ dimensionless variables

$$\left.\begin{aligned} X_1 &= \rho' R_o' v_o' \mu^{-1} = \frac{v_o R_o}{\nu} \\ X_2 &= \rho^0 R_o^{-1} v_o^0 k' = \frac{k}{R_o} \\ X_3 &= \rho^0 R_o^0 v_o^0 S' = S \\ X_4 &= \rho^0 R_o^{-1} v_o^2 g^{-1} = \frac{v_o^2}{g R_o} \end{aligned}\right\} \tag{1.67}$$

Any dimensionless property Π_A of the flow related to a location (point) given by the co-ordinates x, y, z must be a certain function of the dimensionless variables (1.67) and of the dimensionless co-ordinates ξ, η, ζ. If the non-uniform flow under consideration is in addition non-stationary, i.e. if it varies with the time t, then R_o and v_o have to be regarded as the values corresponding to a particular instant, say $t = 0$. In this case, the property Π_A will be dependent also on the dimensionless time θ. The dimensionless co-ordinates ξ, η, ζ and the dimensionless time θ are given by their dimensional counterparts x, y, z and t, and by the basic quantities ρ, R_o and v_o as follows:

$$\left.\begin{aligned} \xi &= \rho^0 R^{-1} v_o^0 x = x/R_o \\ \eta &= \rho^0 R^{-1} v_o^0 y = y/R_o \\ \zeta &= \rho^0 R^{-1} v_o^0 z = z/R_o \\ \theta &= \rho^0 R^{-1} v_o' t = t v_o/R_o \end{aligned}\right\} \tag{1.68}$$

Here the dimensionless time θ which can appear in the dimensionless expression of any time dependent phenomenon (non steady-state phenomenon) is referred to as the Strouhal number. Hence, the dimensionless version Π_A corresponding to a property A of a non-stationary and non-uniform flow can be expressed by the following functional relation:

$$\Pi_A = \varphi_A(\xi, \eta, \zeta, \theta, X_1, X_2, X_3, X_4) \tag{1.69}$$

If the non-uniform flow is stationary, then (1.69) reduces into

$$\Pi_A = \varphi_A(\xi, \eta, \zeta, X_1, X_2, X_3, X_4) \tag{1.70}$$

If, in addition, the property A is a property of a *section* (rather than of a *point*), then it is determined only by the co-ordinate x, that is ξ. Accordingly,

$$\Pi_A = \varphi_A(\xi, X_1, X_2, X_3, X_4) \tag{1.71}$$

Consider the following section properties which correspond to a section $x - x$, and which are thus no longer marked by the subscript o:

$$\frac{\mathrm{d}h}{\mathrm{d}x}; \quad \frac{vR}{\nu}; \quad \frac{k}{R}; \quad \frac{v^2}{gR} \tag{1.72}$$

Since these four properties A_i are dimensionless, their versions Π_{A_i} coincide with themselves

$$\Pi_{A_i} = \rho^0 R_o^0 v_o^0 A_i = A_i \qquad (i = 1, 2, 3, 4) \tag{1.73}$$

and therefore we have for them

$$\left.\begin{aligned}
\frac{\mathrm{d}h}{\mathrm{d}x} &= \varphi_1(\xi, X_1, X_2, X_3, X_4)\\
\frac{vR}{\nu} &= \varphi_2(\xi, X_1, X_2, X_3, X_4)\\
\frac{k}{R} &= \varphi_3(\xi, X_1, X_2, X_3, X_4)\\
\frac{v^2}{gR} &= \varphi_4(\xi, X_1, X_2, X_3, X_4)
\end{aligned}\right\} \tag{1.74}$$

The elimination of three quantities X_1, X_2 and X_4 (containing v_o and/or R_o) from the four independent equations above yields:

$$\frac{\mathrm{d}h}{\mathrm{d}x} = \bar{\varphi}_1\left(\xi, \frac{vR}{\nu}, \frac{k}{R}, S, \frac{v^2}{gR}\right) \tag{1.75}$$

The difference between the first equation (1.74) and eqn. (1.75) lies in the fact that the value of $\mathrm{d}h/\mathrm{d}x$, corresponding to a section $x - x$ is given by the former equation in terms of the characteristics of the section $\mathrm{O} = \mathrm{O}$, whereas by the latter equation it is given in terms of the characteristics of 'its own' section $x - x$.

It is intended now to compare the relation (1.75) supplied by the theory of dimensions with the corresponding formula of hydraulics. The classical formula of hydraulics for a stationary non-uniform flow with a free surface in a non-prismatic channel is derived by assuming that the flow varies (along x) *gradually*. In other words, it is assumed that the streamlines of the flow might possess a certain curvature and divergence (convergence), but this is small. This assumption regarding gradual variation is introduced for the following reasons:

(a) in order to express the pressure distribution, in the cross-sections of the flow, according to the linear law of hydrostatics,
(b) in order to ensure that the rate of energy loss along x is entirely due to friction, and not due to sudden expansion, sudden contraction, and other non-gradual changes in the geometry of the flow.

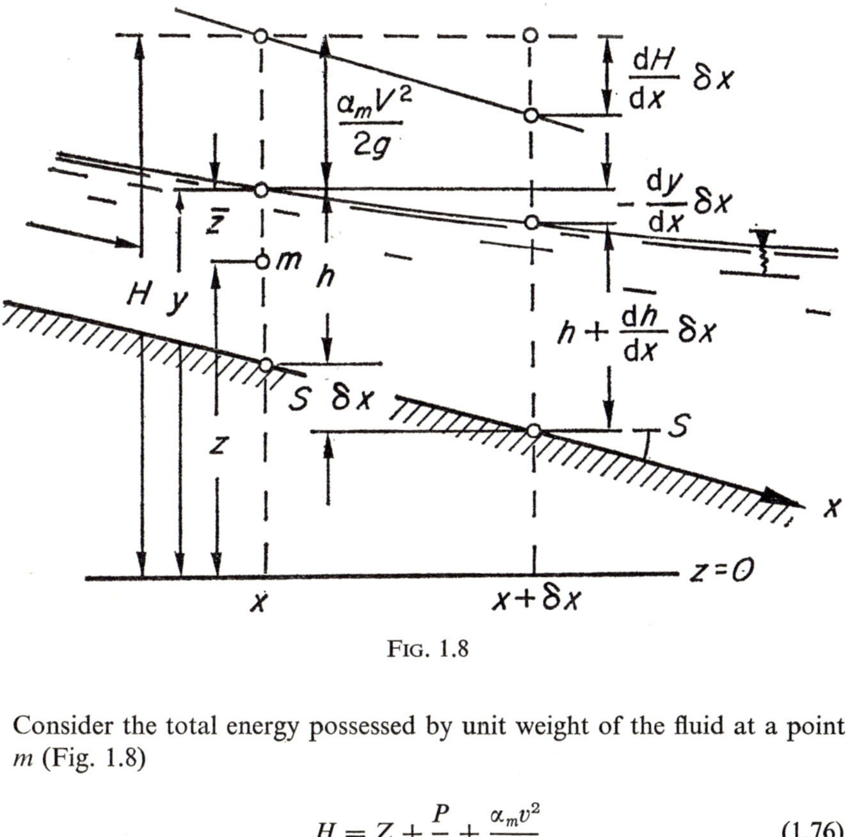

FIG. 1.8

Consider the total energy possessed by unit weight of the fluid at a point m (Fig. 1.8)

$$H = Z + \frac{P}{\gamma} + \frac{\alpha_m v^2}{2g} \tag{1.76}$$

where Z and p are the elevation and pressure at the point m, v the average flow velocity in the section containing the point m, whereas α_m is the coefficient which takes into account the fact that the velocity at the point

m might not necessarily be equal to the velocity v. Since the pressure is supposed to be distributed according to the hydrostatic law, we have

$$p = \gamma\bar{Z} \tag{1.77}$$

and therefore (1.76) can be expressed as follows:

$$H = y + \frac{\alpha_m v^2}{2g} \tag{1.78}$$

where

$$y = Z + \bar{Z} \tag{1.79}$$

is the elevation of the free surface. Observe that in the version (1.78) it is only α_m that reflects the difference from one point of the cross-section to another. Introducing an appropriate average value α (instead of individual values α_m) the relation (1.78) can be written in the following manner

$$H = y + \frac{\alpha v^2}{2g} \tag{1.80}$$

where the value of H is given by the characteristics of the corresponding section (independently of the characteristics of individual points). Differentiating (1.80) by x we arrive at

$$\frac{\mathrm{d}H}{\mathrm{d}x} = \frac{\mathrm{d}}{\mathrm{d}x}\left(y + \frac{\alpha v^2}{2g}\right) = \frac{\mathrm{d}y}{\mathrm{d}x} + \frac{\alpha}{2g}\frac{\mathrm{d}v^2}{\mathrm{d}x} \tag{1.81}$$

Since the total Energy H decreases when x increases (i.e. in the direction of flow) the rate of change $\mathrm{d}H/\mathrm{d}x$ is always negative. Accordingly, the quantity $-\mathrm{d}H/\mathrm{d}x$, which will subsequently be referred to as 'the energy gradient' is always positive. From Fig. 1.8, it is clear that the first term on the right hand side of (1.81) can be expressed as follows:

$$\frac{\mathrm{d}y}{\mathrm{d}x} = \frac{\mathrm{d}h}{\mathrm{d}x} - S \tag{1.82}$$

where S is the slope of the channel bed (which is treated as a positive quantity), while $\mathrm{d}h/\mathrm{d}x$ is the rate of increment of the flow depth. Let us now go over to the consideration of the second term on the right hand side of (1.81). Bearing in mind that the average velocity v is given by

$$v = \frac{Q}{\omega} \tag{1.83}$$

and taking into account that the flow rate Q is supposed to remain constant along x, the second term in (1.84) can be expressed as

$$\frac{\alpha}{2g}\frac{dv^2}{dx} = \frac{\alpha Q^2}{2g}\frac{d\omega^{-2}}{dx} = -\frac{\alpha Q^2}{g\omega^3}\frac{d\omega}{dx} \qquad (1.84)$$

Here the cross-section area ω varies with both x and h. Thus, the total differential of the function of two variables $\omega = \varphi(x, h)$ is

$$d\omega = \frac{\partial\omega}{\partial x}dx + \frac{\partial\omega}{\partial h}dh \qquad (1.85)$$

while its total derivative with respect to x being

$$\frac{d\omega}{dx} = \frac{\partial\omega}{\partial x} + \frac{\partial\omega}{\partial h}\frac{dh}{dx} = \frac{\partial\omega}{\partial x} + B\frac{dh}{dx} \qquad (1.86)$$

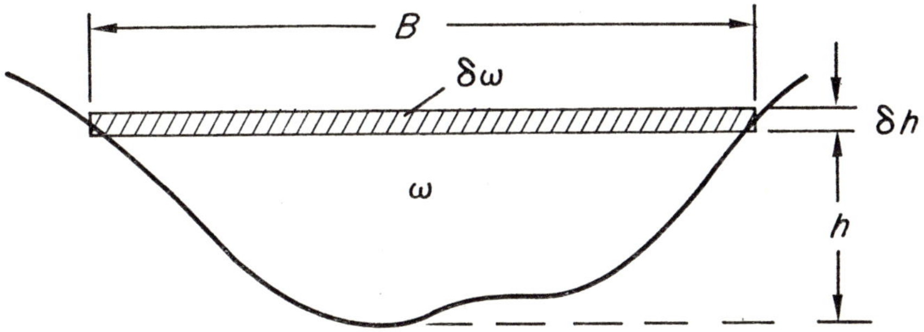

FIG.1.9

for, as is clear from Fig. 1.9, the increment $\delta\omega$ due to the increment δh of the flow depth alone is given by $\delta\omega = B\,\delta h$ and thus

$$\frac{\partial\omega}{\partial h} = B \qquad (1.87)$$

Substituting (1.86) in (1.84) and considering that Q^2/ω^2 implies v^2 while ω is equal to $R\chi$ (where R is the hydraulic radius, while χ the wetted perimeter) we can express the second term as follows:

$$\frac{\alpha}{2g}\frac{dv^2}{dx} = -\alpha\frac{v^2}{gR\chi}\left(\frac{\partial\omega}{\partial x} + B\frac{dh}{dx}\right) \qquad (1.88)$$

Let us now substitute this value of the second term and the value (1.82) of the first term in the eqn. (1.81). We obtain

$$\frac{dH}{dx} = \frac{dh}{dx} - S - \alpha\frac{v^2}{gR}\left(\frac{1}{\chi}\frac{\partial\omega}{\partial x} + \frac{B}{\chi}\frac{dh}{dx}\right) \qquad (1.89)$$

At the present state of knowledge, it is not exactly known how the energy gradient $-\mathrm{d}H/\mathrm{d}x$ of a non-uniform flow is determined by the characteristics of that flow. Hence, in contemporary hydraulics, one has no alternative but to assume that in the case of a gradually varying non-uniform flow (where the energy loss may be regarded as entirely due to the friction (as implied by (b)), the value of $-\mathrm{d}H/\mathrm{d}x$ is given in exactly the same manner as in the case of a uniform flow, i.e. as

$$-\frac{\mathrm{d}H}{\mathrm{d}x}=\frac{v^2}{c^2gR} \tag{1.90}$$

In the case of a uniform flow, the slope S coincides with the energy gradient $-\mathrm{d}H/\mathrm{d}x$, and therefore the eqn. (1.90) is another way of writing eqn. (1.58) for this case.

Substituting this value of $\mathrm{d}H/\mathrm{d}x$ in (1.93) we arrive at

$$\frac{v^2}{c^2gR}=S-\frac{\mathrm{d}h}{\mathrm{d}x}+\alpha\frac{v^2}{gR}\left(\frac{1}{\chi}\frac{\partial\omega}{\partial x}+\frac{B}{\chi}\frac{\mathrm{d}h}{\mathrm{d}x}\right) \tag{1.91}$$

and solving $\mathrm{d}h/\mathrm{d}x$ at

$$\frac{\mathrm{d}h}{\mathrm{d}x}=\frac{S-\dfrac{v^2}{gR}\left(\dfrac{1}{c^2}-\dfrac{\alpha}{\chi}\dfrac{\partial\omega}{\partial x}\right)}{1-\alpha\dfrac{B}{\chi}\dfrac{v^2}{gR}} \tag{1.92}$$

which is a well-known formula in hydraulics.

Even though our objective here is to compare the expression (1.92) with the relation (1.75) determined from the dimensional considerations, we have not simply written the formula (1.92), but have derived it. This is because when studying the similarity of non-uniform flow in open channels and rivers (Chapter 5), we shall frequently refer to the formula (1.92), and then it will be relevant to be acquainted not only with this formula itself, but also with all the conditions which limit its validity.

Let us now compare (1.75) with (1.92). According to (1.92), the value of $\mathrm{d}h/\mathrm{d}x$ is determined by the following dimensionless quantities:

$$S,\quad \frac{v^2}{gR},\quad c,\quad \alpha\frac{B}{\chi}\quad \text{and}\quad \frac{\alpha}{\chi}\frac{\partial\omega}{\partial x}\,. \tag{1.93}$$

Observe that the last two combinations are formed solely by the geometric characteristics of the flow cross-section. Hence, their numerical values can vary only if the position of the cross-section along x varies, and thus the consideration of these geometric combinations can always be substituted by that of $\xi = x/R$. Thus, taking into account that the friction factor c is but a symbol which stands for a certain function of the Reynolds number vR/ν and the relative roughness k/R, it becomes obvious that the consideration of five dimensionless quantities (1.93) is entirely equivalent to the consideration of the following five quantities:

$$S, \frac{v^2}{gR}, \frac{vR}{\nu}; \frac{k}{R} \quad \text{and} \quad \xi \tag{1.94}$$

which are simply the variables of the dimensionless form (1.75) determined from purely dimensional considerations.

It might be worthwhile to mention that if the non-uniform flow is not gradually varying, then the energy loss per unit length of the flow, that is $\mathrm{d}H/\mathrm{d}x$, cannot be entirely due to friction. Accordingly, the right hand side of (1.90) cannot consist solely of the term v^2/c^2gR reflecting the influence of the friction factor c alone; some other terms taking into account the losses caused by *non-gradual* variations in the flow geometry must also be included. In other words, in the case of a non-gradually varying, non-uniform flow, the formula for $\mathrm{d}h/\mathrm{d}x$ must be more complicated than (1.92), not to mention the complications that will arise in the mathematical forms due to the non-linear pressure distribution. However, one must not infer from this that the set of dimensionless variables (1.94) is valid for gradually varying non-uniform flows only. The mentioned complication in the expression of $\mathrm{d}h/\mathrm{d}x$ can arise only due to some additional terms reflecting the influence of non-gradual changes in the geometry, thus, these terms cannot be anything but functions of various geometric parameters. Since, in the derivation of the dimensionless form (1.75) the geometry of the flow (however it might be) was assumed specified, the variables which appear in that form (i.e. the dimensionless variables (1.94)) must be valid independently of whether the non-uniform flow varies gradually or not.

For the further study of the subject presented in this and in the next chapter, the treatises shown in the bibliography at the end of this chapter are recommended.

REFERENCES

1. L. I. Sedov, *Similarity and Dimensional Methods in Mechanics*, Academic Press (1959).
2. G. Birkhoff, *Hydrodynamics, a Study in Logic, Fact and Similitude*, Harvard University Press (1950). Also Dover Publications (1955).
3. P. W. Bridgman, *Dimensional Analysis*, Yale University Press (1931).
4. J. E. Plapp, *Engineering Fluid Mechanics*, Chapter 6: Similarity and Dimensional Analysis of Fluid Flows, Prentice-Hall (1968).
5. J. Palacios, *Dimensional Analysis*, Macmillan, London (1964).
6. H. L. Langhaar, *Dimensional Analysis and Theory of Models*, John Wiley (1962).

7. G. Murphy, *Similitude in Engineering*, The Ronald Press Co. (1950).
8. R. Comolet, *Introduction à L'Analyse Dimensionelle et aux Problèms de Similitude en Méchanique des Fluides*, Masson et Cie, Paris (1958).
9. D. C. Ipsen, *Units, Dimensions, and Dimensionless Numbers*, McGraw-Hill (1960).
10 R. C. Pankhurst, *Dimensional Analysis and Scale Factors*, Chapmann and Hall Ltd., London. Reinhold Publishing Corp., New York (1964).
11. M. S. Yalin, Über die Bedeutung, der Theorie der Dimensionen für das wasserbauliche Versuchwesen, *Die Bautechnik*, **8**, August 1959.
12. S. J. Kline, *Similitude and Approximation Theory*, McGraw-Hill (1965).

2 Principles of the Theory of Similarity

2.1 The Idea of a Model

Since the numerical values of the dimensionless quantities remain the same in all systems of fundamental units, the numerical values of the dimensionless quantities X_j and Π_A, encountered in the preceding chapter, will not change if the system of fundamental units L', T' and M' by which they are expressed is replaced by a different system of units L'', T'' and M''. Two systems of fundamental units can always be related to each other by the 'transformation formulae'

$$\begin{aligned} L'' &= \lambda_L L' \\ T'' &= \lambda_T T' \\ M'' &= \lambda_M M' \end{aligned} \tag{2.1}$$

where the proportionality factors λ_L, λ_T and λ_M are some arbitrary constants. Accordingly, the dimensionless numbers in general (and X_j and Π_A in particular) can be regarded as the quantities which remain invariant with respect to the transformation (2.1).

Consider now the first proportionality in (2.1) which can be expressed as follows:

$$\lambda_L = \frac{L''}{L'}$$

This is the ratio of two different length units, or the ratio of two numerical values determined for the *same* length measured by two *different* units. On the other hand, the above ratio can equally well be interpreted as the ratio of two different lengths measured in the same units. Similarly

$$\lambda_T = \frac{T''}{T'}$$

can be regarded as the ratio of two different time intervals (both measured in the same time units) rather than the ratio of two different values (given by two different units) for the same duration. It is clear that the same argument is true for the third proportionality of (2.1).

From this new interpretation of (2.1) it follows that the numerical values of all the dimensionless quantities X_j and Π_A, and thus of all dimensionless functional relations

$$\Pi_A = \varphi_A(X_1, X_2, \ldots X_N)$$

representing the laws governing mechanical phenomena, would remain exactly the same if the fundamental entities—length, time and mass—were to become λ_L, λ_T and λ_M times different. No change in the progress of any mechanical, or even physical, phenomenon could normally be detected if the size of our world were to become λ_L times smaller, if time were to flow λ_T times faster, or if the mass of every substance were to become λ_M times, denser.

The above explanations brings us immediately to the idea that a mechanical phenomenon can be studied in an artificially made 'small world', referred to as a *scale model*, where all lengths, times and masses are scaled down λ_L, λ_T and λ_M times, respectively.

2.2 Definition of Dynamic Similarity

(i) The systems S' and S'' related to each other by the first proportionality of (2.1)

$$\lambda_L = \frac{L''}{L'} \tag{2.2}$$

are referred to as *geometrically similar systems.* Any two geometrically similar systems can be brought into a parallel (homologous) position as in Fig. 2.1, where the straight lines passing through the corresponding points a'; a'', b'; b'' etc. are converging at the same point O (referred to as the homology centre). It is clear that in the proportionality above, L' and L'' represent any two corresponding lengths of the systems S' and S'', as implied by the new interpretation of the set (2.1) (used in the theory of similarity). L' and L'' should no longer be interpreted as two different length units (as is done in the theory of dimensions).

(ii) The systems S' and S'' which are related to each other by the first two proportionalities of (2.1)

$$\lambda_L = \frac{L''}{L'};\ \lambda_T = \frac{T''}{T'} \tag{2.3}$$

are referred to as *kinematically similar systems*. Here L', T' and L'', T'', are corresponding lengths and times of the systems S' and S'' respectively. Let B' and B'' be two corresponding geometrically similar fluid elements (fluid balls) which are at the instant $t = 0$ in the corresponding positions O′ and O″. The kinematic similarity of the systems S' and S'' implies that if the moving fluid element B' at the instants t_1', t_2', t_3', . . . t_n', passes

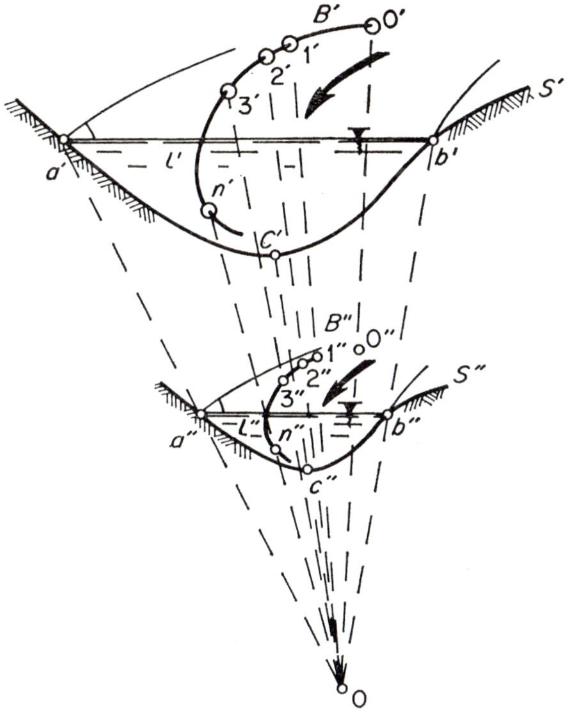

FIG. 2.1

through the points 1′, 2′, 3′, . . . n' (forming the path l' of the fluid element, B'), then the corresponding fluid element B'', at the instants t_1'', t_2'', t_3'', . . . t_n'', passes through the corresponding points 1″, 2″, 3″, . . . n'' (forming a path l'' that is geometrically similar to l') in such a manner that the ratio between the corresponding time intervals is constant, i.e.

$$\frac{t_2'' - t_1''}{t_2' - t_1'} = \frac{t_3'' - t_2''}{t_3' - t_2'} = \ldots \frac{t_n'' - t_{n-1}''}{t_n' - t_{n-1}'} = \lambda_T = \text{const.}$$

(iii) The systems S' and S'' related to each other by all three proportionalities of (2.1)

$$\lambda_L = \frac{L''}{L'}; \quad \lambda_T = \frac{T''}{T'}; \quad \lambda_M = \frac{M''}{M'} \tag{2.4}$$

are referred to as *dynamically similar systems.* Dynamically similar systems are distinguished from kinematically similar systems by the

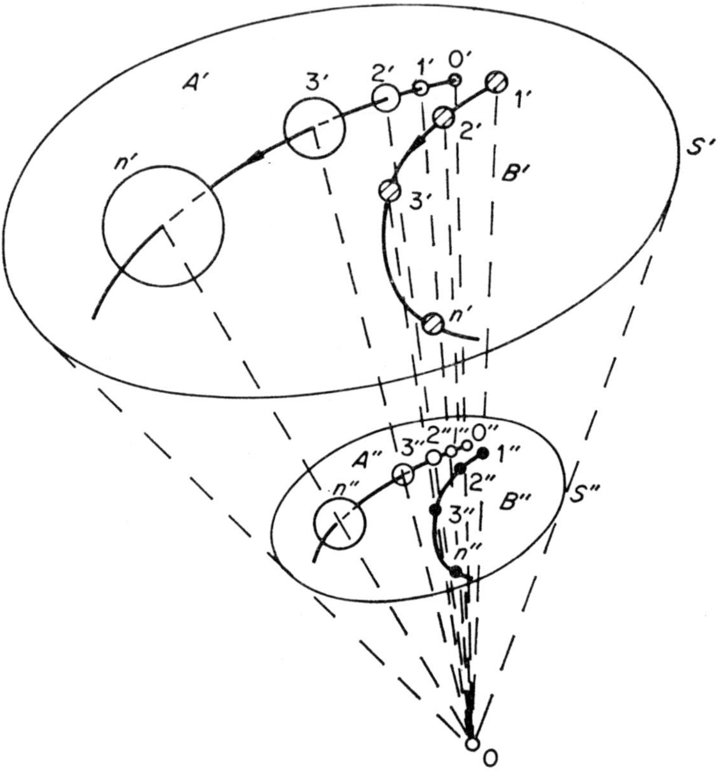

FIG. 2.2

following additional condition. The masses M' and M'' of all corresponding (geometrically similar) elements, in all corresponding times, must be related to each other by the same constant proportion λ_M. Consider, for example, the 'expanding' elements A' and A'' shown in Fig. 2.2. The masses $M_A{}'$ and $M_A{}''$ of these elements moving along the geometrically similar paths vary as the functions of the positions $1', 2', 3', \ldots n'$ and $1'', 2'', 3'', \ldots n''$ (or as the functions of the instants $t_1{}', t_2{}', \ldots t_n{}'$ and

$t_1'', t_2'', \ldots t_n''$). Dynamic similarity of the systems S' and S'' requires that the condition

$$\frac{(M_A)_1''}{(M_A)_1'} = \frac{(M_A)_2''}{(M_A)_2'} = \ldots \frac{(M_A)_n''}{(M_A)_n'} = \lambda_M = \text{const}$$

is satisfied. Let us assume that the system S' involves another element B' of a say, 'non-expanding' nature but which is of an entirely different density from the element A'. Thus, the mass M_B' of the element B' might never be equal to that of A', i.e. the following condition can be present

$$M_B' \neq (M_A)_{i'}' \text{ for any position } i' \qquad \text{(or instant } t')$$

In this case, the similar condition must be valid for the corresponding element B'' of the system S''

$$M_B'' \neq (M_A)_{i''}'' \text{ for any position } i'' \qquad \text{(or instant } t'')$$

If the systems S' and S'' are dynamically similar, then, despite these inequalities between the masses of the elements B and A, the ratio M_B''/M_B' must be identical to the same constant value λ_M as the ratio $(M_A)_{i''}''/(M_A)_{i'}'$, i.e.

$$\frac{M_B''}{M_B'} = \frac{(M_A)_{i''}''}{(M_A)_{i'}'} = \lambda_M$$

must be valid for all corresponding positions i' and i''.

2.3 Dynamically Similar Models and their Scales

In the following, the system S' will be identified with the natural phenomenon or the *prototype*, while the system S'' will be identified with its *model.* The constant proportionality coefficients λ_L, λ_T and λ_M will be referred to as the geometric scale (or simply the model scale), the time scale, and the mass scale respectively. Let a be any quantity:

the prototype value of a will be denoted by a'
the model value of a will be denoted by a''
the scale of a will be denoted by $\lambda_a = a''/a'$

At the beginning of this chapter it was pointed out that the dimensionless laws governing mechanical phenomena such as

$$\Pi_A = \varphi_A(X_1, X_2, \ldots X_n)$$

remain unchanged with respect to the transformation set (2.1). Subsequently, we have seen that the very same set defines the dynamic similarity (eqn. (2.4)). Hence, the dynamic similarity implies the identity of all the dimensionless laws governing the model and prototype phenomena. Note, that in order to achieve dynamic similarity it is sufficient to provide constant values for the scales of only three independent quantities (length, time, and mass), yet by having succeeded in this one automatically achieves the identity of all dimensionless relations. This is why in a (truly) dynamically similar model one can study any property of the phenomenon under investigation. If in the handy 'minature world' called 'model', the model value A'' of a property A is determined, then the prototype value A' can be predicted simply by multiplying the model value A'' by the scale of this property λ_A, that is, as

$$A' = \lambda_A A''$$

The model technique is especially valuable if the direct measurement of the prototype value A' is not only inconvenient but impossible (e.g. if A' is a property of a future structure). The following chapters of the present book deal with the reproduction of various hydraulic phenomena on small scale models. However, before going into this, it is necessary to become acquainted with certain notions which are relevant to the clearer understanding of the study.

(i) The unit (dimension) of a quantity a' expressed in terms of the fundamental units L', T' and M' is given by

$$[a]' = L'^{\alpha} T'^{\beta} M'^{\gamma}$$

The unit of the same quantity, expressed in terms of the fundamental units L'', T'' and M'' being

$$[a]'' = L''^{\alpha} T''^{\beta} M''^{\gamma}$$

Thus

$$\frac{[a]''}{[a]'} = \left(\frac{L''}{L'}\right)^{\alpha} \left(\frac{T''}{T'}\right)^{\beta} \left(\frac{M''}{M'}\right)^{\gamma}$$

Interpreting the ratio of two different units as the ratio of two different magnitudes (i.e. as the scale) we arrive at the following expression for the scale λ_a of the quantity a

$$\lambda_a = \lambda_L{}^{\alpha} \lambda_T{}^{\beta} \lambda_M{}^{\gamma} \tag{2.5}$$

Hence, the scale λ_a of a dimensional quantity a can be obtained from the expression of its dimension simply by substituting L, T, M and $[a]$ by λ_L, λ_T, λ_M and λ_a. If a is a dimensionless quantity ($\alpha = \beta = \gamma = 0$), then the relation (2.5) gives

$$\lambda_a = \lambda_L{}^0 \lambda_T{}^0 \lambda_M{}^0 = 1 \tag{2.6}$$

which is a different way of saying that in dynamically similar systems all the corresponding dimensionless quantities are identical.

(ii) Take any three dimensional quantities a_1, a_2 and a_3 which possess independent dimensions. According to (2.5) we have

$$\left.\begin{aligned} \lambda_{a_1} &= \lambda_L^{\alpha_1}\,\lambda_T^{\beta_1}\,\lambda_M^{\gamma_1} \\ \lambda_{a_2} &= \lambda_L^{\alpha_2}\,\lambda_T^{\beta_2}\,\lambda_M^{\gamma_2} \\ \lambda_{a_3} &= \lambda_L^{\alpha_3}\,\lambda_T^{\beta_3}\,\lambda_M^{\gamma_3} \end{aligned}\right\} \tag{2.7}$$

Since a_1, a_2 and a_3 are of independent dimensions and thus since

$$\Delta = \begin{vmatrix} \alpha_1 & \beta_1 & \gamma_1 \\ \alpha_2 & \beta_2 & \gamma_2 \\ \alpha_3 & \beta_3 & \gamma_3 \end{vmatrix} \neq 0$$

the system (2.7) possesses an unequivocal solution with respect to λ_L, λ_T and λ_M which can be expressed as follows

$$\left.\begin{aligned} \lambda_L &= \lambda_{a_1}^{k_1}\,\lambda_{a_2}^{l_1}\,\lambda_{a_3}^{m_1} \\ \lambda_T &= \lambda_{a_1}^{k_2}\,\lambda_{a_2}^{l_2}\,\lambda_{a_3}^{m_2} \\ \lambda_M &= \lambda_{a_1}^{k_3}\,\lambda_{a_2}^{l_3}\,\lambda_{a_3}^{m_3} \end{aligned}\right\} \tag{2.8}$$

Hence, the dynamic similarity, which has been formally defined in Section 2.2 by the scales of length, time, and mass, can in fact be defined by the scales of *any* three quantities possessing independent dimensions, for a 'trio' λ_{a_1}, λ_{a_2} and λ_{a_3} defines unequivocally the 'trio' λ_L, λ_T and λ_M (used in the formal definition).

Since the dimension of any quantity a can equally well be defined by the units of any three dimensionally independent quantities a_1, a_2 and a_3 (see eqn. (1.10)), it follows that the scale of the quantity a can be given also as

$$\lambda_a = \lambda_{a_1}^{\bar{\alpha}}\,\lambda_{a_2}^{\bar{\beta}}\,\lambda_{a_3}^{\bar{\gamma}} \tag{2.9}$$

(iii) At the beginning of this section it has been pointed out that if the systems S' and S'' are dynamically similar, then all the corresponding dimensionless relations of the systems S' and S'' are identical. One can show the converse statement is also valid, that is, that the identity of all the dimensionless relations implies the dynamic similarity. But, for a given geometry, the values of all the dimensionless properties Π_A of a phenomenon are completely determined by only $N = n - 3$ dimensionless combinations X_j. Thus, if the systems are geometrically similar, then the identity of all the dimensionless properties, and thus the dynamic similarity, will automatically be achieved if the identity of only $N = n - 3$ dimensionless combinations X_j is provided. Hence, in order to achieve dynamic similarity of a phenomenon it is necessary and sufficient to provide the identity of the dimensionless variables X_j, i.e. to provide the validity of the scale relations

$$\lambda_{X_1} = 1; \lambda_{X_2} = 1; \ldots \lambda_{X_N} = 1 \qquad (N = n - 3) \tag{2.10}$$

and to ensure the existence of geometric similarity. The existence of geometric similarity is important because the form of a function such as

$$\Pi_A = \varphi_A(X_1, X_2, \ldots X_N)$$

depends entirely on the geometry. Accordingly, if the systems S' and S'' are not geometrically similar, then one cannot be sure that the form of the function φ_A is the same, in model and prototype, and consequently one cannot claim that the identity of the dimensionless variables X_j will yield the identity of the dimensionless properties Π_A. Considering (1.19), the scale relations (2.10) can be expressed in the following (open) form

$$\left.\begin{aligned}
\lambda_{X_1} &= \lambda_{a_1}{}^{x_1}\,\lambda_{a_2}{}^{y_1}\,\lambda_{a_3}{}^{z_1}\,\lambda_{a_4}{}^{m_1} = 1\\
\lambda_{X_2} &= \lambda_{a_1}{}^{x_2}\,\lambda_{a_2}{}^{y_2}\,\lambda_{a_3}{}^{z_2}\,\lambda_{a_5}{}^{m_2} = 1\\
&\cdots\cdots\cdots\cdots\cdots\cdots\\
\lambda_{X_N} &= \lambda_{a_1}{}^{x_N}\,\lambda_{a_2}{}^{y_N}\,\lambda_{a_3}{}^{z_N}\,\lambda_{a_N}{}^{m_N} = 1
\end{aligned}\right\} \tag{2.11}$$

or

$$\left.\begin{aligned}
\lambda_{a_4}{}^{m_1} &= \lambda_{a_1}{}^{-x_1}\,\lambda_{a_2}{}^{-y_1}\,\lambda_{a_3}{}^{-z_1}\\
\lambda_{a_5}{}^{m_2} &= \lambda_{a_1}{}^{-x_2}\,\lambda_{a_2}{}^{-y_2}\,\lambda_{a_3}{}^{-z_2}\\
&\cdots\cdots\cdots\cdots\cdots\cdots\\
\lambda_{a_N}{}^{m_N} &= \lambda_{a_1}{}^{-x_N}\,\lambda_{a_2}{}^{-y_N}\,\lambda_{a_3}{}^{-z_N}
\end{aligned}\right\} \tag{2.12}$$

The scale relationships constituting the systems (2.11) or (2.12) are referred to as the *criteria of similarity* of the phenomenon given by n characteristic parameters a_i. In an analogous manner, using the equation (1.20) in the relation $\lambda_{\Pi_A} = 1$, we determine the value of the scale λ_A of a property A as

$$\lambda_A = \lambda_{a_1}{}^{-x_A}\,\lambda_{a_2}{}^{-y_A}\,\lambda_{a_3}{}^{-z_A} \tag{2.13}$$

It should be remembered that the basic quantities a_1, a_2 and a_3 can be any three parameters having independent dimensions. Thus, by giving the scales λ_{a_i} and λ_A in terms of the scales of basic quantities, in fact we are giving them in terms of the scales of any three dimensionally independent characteristic parameters.

It follows that the dynamic similarity of a prototype phenomenon (system S') in a small scale model (system S'') must be achieved according to the following procedure:

1. Out of n characteristic parameters a_i of the phenomenon under investigation we select (theoretically at random) three dimensionally

independent parameters, say a_1, a_2 and a_3, and their scales λ_{a_1}, λ_{a_2} and λ_{a_3}.

2. Using the selected values λ_{a_1}, λ_{a_2} and λ_{a_3} we determine from (2.12) the values of the remaining scales λ_{a_4}, λ_{a_5}, . . . λ_{a_N}.
3. Knowing, thus, the scales λ_{a_1} of all n characteristic parameters a_i, and knowing the prototype values a_i' of these characteristic parameters we determine the values of n model parameters a_i'' as

$$a_i'' = \lambda_{a_i} a_i' \qquad (i = 1, 2, \ldots n)$$

and set them on the geometrically similar model. Such required values of a_i'' in the model must always be available, in principle, since the characteristic parameters are independent quantities.

The model (system S'') designed according to the procedure described above is referred to as the *dynamically similar* model of its prototype (system S'). Any property A' of the prototype can be predicted from a dynamically similar model, by measuring the model value A'' and by multiplying it by the scale λ_A.

Attention is once more drawn to the fact that in order to establish the criteria of dynamic similarity (eqns. (2.11) and (2.12)) of a mechanical phenomenon, it is necessary and sufficient to know the set of n characteristic parameters of that phenomenon. If the set of characteristic parameters is revealed, then the determination of the set of dimensionless variables, and consequently of the criteria of dynamic similarity, is a matter of application of exact and simple rules. This fact is emphasized because, in spite of its perpetual demonstration for more than three decades in the special literature on the theory of dimensions and similarity, in the majority of textbooks on hydraulics the criteria of similarity (Froude, Reynolds, Webber, Strouhal numbers, etc.) are still persistently derived either from the expressions of various individual forces or from certain differential equations (such as Navier Stokes eqns.) representing the combinations of these forces. Hence, a person acquainted with the notions of hydrodynamic similarity from the traditional textbooks might well be under the impression that the criteria of similarity can only follow from certain equations reflecting the physical nature of a phenomenon and he might be sceptical about this new idea of determining the similarity criteria from a certain number of discrete symbols. However, as has been mentioned above, the idea (which unfortunately does not usually appear in textbooks on hydraulics) is not new, while the fact that the criteria of similarity follow from the parameters themselves rather than from a relation (an equation) among them, forms the most advantageous feature of the dimensional method. Indeed, it is far simpler (and safer) to determine the parameters themselves rather than to determine, in addition, the relations among these parameters. For example, at present we know perfectly well what are the characteristic parameters of a flow taking place on an erodible granular medium (bed), and therefore we also know perfectly well what are the criteria of similarity of sediment transport.* Yet, if we were to attempt to establish these criteria from the equations of sediment transport we would run into serious difficulties, for in the present state of knowledge not a single equation of sediment transport can be regarded as known (in the rigorous sense of the word). In the chapters which follow we shall see that the criteria of similarity derived for various phenomena of hydraulics cannot usually be applied easily in conventional models where the prototype fluid (water) is used. But this has no bearing on '*what* these criteria are'; it is the matter of '*how* they should be utilized'.

2.4 Hydraulic Models

In the case of hydraulic models used in practice, the selection of the basic scales λ_{a_1}, λ_{a_2} and λ_{a_3} is frequently subjected to severe restrictions of a physical, technical and/or economic nature. For example, for technical and economic reasons, in conventional hydraulic models the prototype fluid (water) is used. This means that when determining the scales of a model, in accordance with the procedure explained in the previous section, we are compelled to 'select'

$$\lambda_\mu = 1 \quad \text{and} \quad \lambda_\rho = 1 \tag{2.14}$$

and thus it is no longer three but only one scale that can be freely selected (for example, the geometric model scale λl). If the acceleration due to gravity g is one of the characteristic parameters, then, since both model and prototype are on the same planet, we have no alternative but to 'choose'

$$\lambda_g = 1 \tag{2.15}$$

which is a physical restriction. Let us now suppose that the model operates with the prototype fluid (water), and that g is one of the characteristic parameters. In this case we have

$$\lambda_\mu = 1; \lambda_\rho = 1; \lambda_g = 1 \tag{2.16}$$

Observe that the quantities μ, ρ and g have independent dimensions, and thus their scales can be identified with the scales λ_{a_1}, λ_{a_2} and λ_{a_3} (which

* See Chapter 6.

appear in the scale relations (2.11) to (2.13)). But from (2.11) to (2.13) it follows clearly that if the scales λ_{a_1}, λ_{a_2} and λ_{a_3} are equal to unity, then all scales are equal to unity. Hence, the validity of (2.16) implies that the model is identical to the prototype or that the realization of a small scale dynamically similar model is impossible. Here we have deliberately avoided the usual reference to the Froude and Reynolds numbers in order to explain the impossibility of the realization of a small scale model. There is no need for μ, ρ and g to be necessarily combined (by means of velocity and length) into the numbers mentioned, not only because we are free in the arrangement of the dimensionless combinations, but also because in the set of characteristic parameters the flow velocity might not even be present (as was the case, for example, in the set of characteristic parameters (1.22) defining the weir flow). In the latter case, there is no possibility of forming the traditional Froude and Reynolds numbers at all, and yet the mere presence of μ, ρ and g in the set of characteristic parameters, and the validity of (2.16) is sufficient to yield unity for all the remaining scales. Hence, the usual demonstration of the impossibility of a small scale model by employing the Froude and Reynolds numbers cannot be regarded as exhaustive.

The previous considerations illustrate how some practical restrictions can either impede, or simply prevent, the realization of a dynamically similar small scale hydraulic model. What should be done in such cases? Or what should be the next best thing to do if the realization of a dynamically similar model becomes impossible? Unfortunately, one cannot give a list of rules on how to achieve the best possible solution. Indeed, each particular case has its own difficulties and thus its own best solution, and it is up to the knowledge, experience, and even intuition of the experimenter to see it. On the other hand, all these different individual solutions are determined as the answers to the following question, common to all similarity problems. 'Knowing that the realization of the dynamically similar behaviour of *all* properties is impossible, how to ensure the dynamical similarity of at least those properties which will be measured and/or observed?' This reduction from *all* properties to only *some* gives rise to certain possibilities which form the subject of the next part of this section.

(i) The reader will recall that n characteristic parameters a_i, and thus $N = n - 3$ dimensionless variables X_j are necessary and sufficient as far as the definition of any property A of the phenomenon is concerned; with respect to a particular property, however, they have to be regarded merely as sufficient. For example, the properties in the midst of a turbulent flow are independent of viscosity (even if the flow boundaries are smooth).

Thus, if the purpose of the model tests is to reveal the behaviour of some properties A of a turbulent flow that are not related to the vicinity of the flow boundaries, then the parameter μ can be excluded from consideration. In this case, even if the model operates with the prototype fluid ($\lambda_\mu = 1$; $\lambda_\rho = 1$) while g is one of the characteristic parameters, we have to take into account only two conditions

$$\lambda_\rho = 1 \quad \text{and} \quad \lambda_g = 1 \tag{2.17}$$

Let λ_ρ and λ_g be λ_{a_1} and λ_{a_2}. Since the parameter μ is excluded and thus the basic quantity a_3 must not be identified with μ any longer, the scale λ_{a_3} must not necessarily be unity. Accordingly, the scales (2.12) of the remaining characteristic parameters and the scales of the properties A, namely

$$\lambda_A = \lambda_{a_1}{}^{-x_A} \lambda_{a_2}{}^{-y_A} \lambda_{a_3}{}^{-z_A} \tag{2.18}$$

need not necessarily be unity, although the product of the first two multipliers on the right hand side of the equation is unity, according to (2.17). The dimensions of ρ, g and a length, say l, are independent, and thus in the relation above, λ_{a_3} can be identified with the scale λ_l of a characteristic length of the phenomenon. Since the scale $\lambda_{a_3} = \lambda_l$ (the geometric model scale) is not subjected to any restriction, the realization of $\lambda_l < 1$, that is, of a small scale model, becomes possible. One should remember, however, that such a model is not a dynamically similar model, in the strict sense of the word, where all the properties of a phenomenon are correctly reproduced. It is a *partly dynamically similar* model where only some (a certain group) of the properties are reproduced adequately. Accordingly, such a model should never be used for the prediction of the behaviour of properties, other than in the group.

These considerations illustrate how the exclusion of a certain parameter a_i, and thus of its scale λ_{o_i} which must satisfy an imposed condition, can provide freedom in the determination of the remaining scales. Sometimes the scale λ_{a_i}, which can be excluded, is not subjected to any special condition (such as $\lambda_{a_i} = 1$) and thus its exclusion does not make any difference with respect to the determination of the remaining scales. But, the mere fact that a parameter *is* excluded, i.e. that it is left outside the 'circle of calculation' provides freedom in the selection of its own scale which might be used in order to achieve some technical or economical advantages. Let us consider, for example, the influence of the flow with B of a river or channel having a trapezoidal cross-section. If the width–depth ratio B/h of the channel is very large, then, as will be seen in Chapter 5, the mechanical properties A in the central regions of such flows can hardly be regarded as dependent on the ratio B/h, Fig. 2.3. Thus, if we are interested in the behaviour of the flow properties, solely in the central regions, we do not need to achieve the identity of the width–depth ratio

$$\lambda_{B/h} = 1 \tag{2.19}$$

and thus the condition

$$\lambda_B = \lambda_h \tag{2.20}$$

which follows from this identity can be relaxed. Let us assume that a certain value of the scale λ_h (which can be identified with λ_{a_3}) is selected. The relaxation of (2.20) implies that the scale λ_B is no longer subjected to any condition, and thus its value can be selected so as to be, for example, smaller than λ_h. This can provide a considerable economy in the model

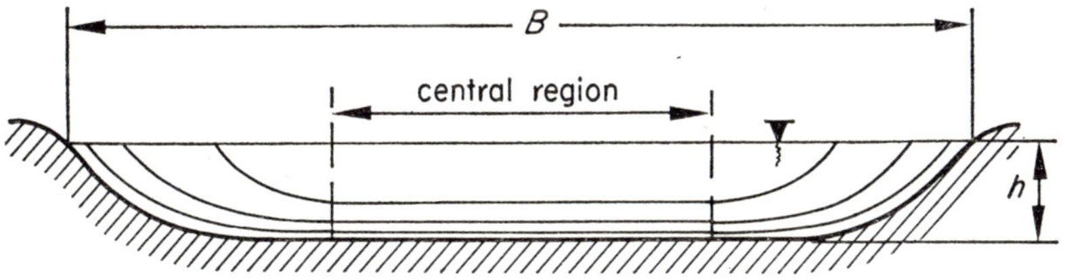

FIG. 2.3

area, and thus in its building and operation costs. Any of the so-called distorted river models (which will be dealt with in Chapter 5) belong to the category of partly dynamically similar models corresponding to the inequality

$$\lambda_B < \lambda_h$$

Here again one must remember that in the model described above only the properties of the central region are reproduced in a reliable manner. Thus, such a model should not be used, for example, in order to predict the behaviour of the flow in the vicinity of the banks. This important limitation which applies to every distorted model ($\lambda_{B/h} \neq 1$) is sometimes not mentioned in model test reports. As a consequence, the client is often under the impression that the behaviour of the prototype phenomenon is predicted (from the distorted model) with the same accuracy for any location in the flow cross-section.

(ii) In some cases it might be of interest to investigate the behaviour of a single property corresponding to a single stage of the phenomenon. For example, when designing a weir all that we might wish to know is whether the downstream part of the structure is adequately protected against the scour (a single property) at the most dangerous flow rate (certain stage). When dealing with ship mooring problems we are generally interested only in the mooring forces acting in rough seas; while when designing an earth dam our concern might be solely with the distribution of filtration velocities in the dam body, which corresponds to a particular (usually maximum) difference of the up and down stream water levels.

If a mathematical relation (formula) involving the property in question is known, then using this relation one can design a small scale model without any need to neglect the influence of any of the characteristic parameters. However, such a small scale model will have the disadvantage of being useful for the investigation of a single property corresponding to a single stage only (or to a very narrow range of the stages).

Let $X_1, X_2, \ldots X_N$ be the dimensionless variables of the phenomenon that possess the quantity A under investigation as a property. Accordingly, the known mathematical relation involving A can be expressed as follows:

$$\Pi_A = \varphi_A(X_1, X_2, \ldots X_N)$$

The dynamically similar reproduction of A implies that

$$\lambda_{\Pi_A} = 1$$

or that

$$\frac{\varphi_A(X_1'', X_2'', \ldots X_N'')}{\varphi_A(X_1', X_2', \ldots X_N')} = 1 \qquad (2.22)$$

is valid. Since

$$X_j'' = \lambda_{X_j} X'$$

is valid, the eqn. (2.22) can be brought into the following relation containing only the scales λ_{X_j} and the prototype values X_j'

$$\frac{\varphi_A(\lambda_{X_1}'X_1', \lambda_{X_2}'X_2', \ldots \lambda_{X_N}X_N')}{\varphi_A(X_1', X_2', \ldots X_N')} = 1 \qquad (2.23)$$

Since the form φ_A is supposed to be known, the mathematical structure of (2.22) is also known. Thus, for a given stage of the prototype phenomenon, i.e. for the given values of the dimensionless variables $X_1', X_2', \ldots X_N'$, the scales $\lambda_{X_1}, \lambda_{X_2}, \ldots \lambda_{X_N}$ can be selected so as to satisfy (2.23). In a dynamically similar model, we have $\lambda_{X_j} = 1$ with respect to any X_j. Thus, the scale relation (2.23) turns into an *identity* that is valid for any form of the function φ_A (i.e. for any property of the phenomenon) and for any set of the values X_j' (i.e. for any stage of the phenomenon). In the chapters that follow, the application of this method will frequently be seen in various examples.

The reader might wonder why we should need a model at all if the mathematical formula for a property A is known, and thus if the value of A can simply be calculated from this formula. The reason for this lies in the difference between knowing a mathematical relation and using it. For example, it is well known that the velocity field of any laminar flow is completely determined by the Navier Stokes equations and the equation of continuity. In spite of this, we should confront an extremely difficult mathematical problem if we attempted to determine from these equations the velocity distribution in practically any flow that is not 'a steady and uniform flow between infinite parallel boundaries'. This would be an

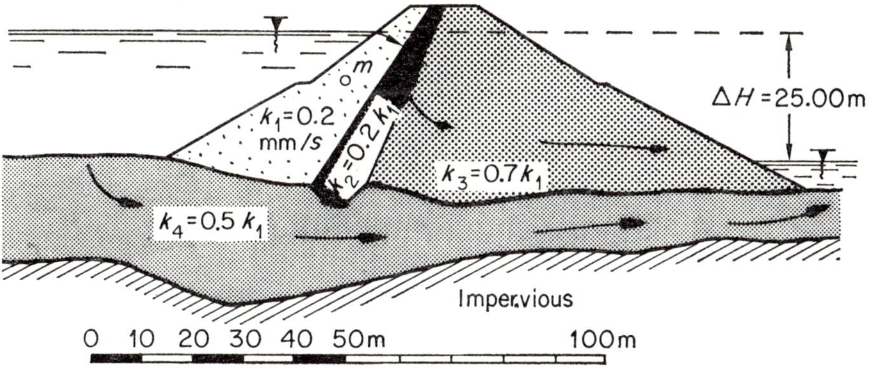

FIG. 2.4

example of the difficulties which arise because the form of the equations is complicated. Sometimes the equations are simple, but the (natural) flow boundaries might be so irregular that one might not be able to formulate them in the language of mathematics. Consider, for example, the filtration flow through a system shown in Fig. 2.4. The value of the filtration velocity v corresponding to any location in the porous medium, is given by as simple a formula as

$$v = kJ \text{ (Darcy law)}$$

where J is the energy gradient, while k is the permeability coefficient. All the geometric dimensions are given (for the figure is accompanied by a scale) the values of the permeability coefficients are also given. What is the value of v at a point, say, marked by m? The answer to this question is not simple at all. On the other hand, using the Darcy relation above, one can design and build (as will be seen in Chapter 4) a small scale hydraulic or electroanalogical model which will give the values of the (single) property v, corresponding to the given stage $\Delta H = 25{\cdot}00$, not only for a particular point m, but for all points of the porous medium.

'Theory of similarity' and 'theory of dimensions' are usually presented together. Hence, for further study on the content of the present chapter, the reader is referred again to the bibliography at the end of Chapter 1.

3 Flows Without Free Surface, Reynolds Models

3.1 General

The stationary flows which belong to this category are essentially those shown schematically in Fig. 1.5a to 1.5h of the first chapter. These flows are either limited by rigid flow boundaries or they are supposed to extend to infinity. We have seen that any dimensionless property Π_A of the flows shown in Fig. 1.5a to 1.5f can be expressed as follows:

$$\Pi_A = \varphi_A(\text{Re}, \Gamma) \tag{3.1}$$

where Π_A and Re are

$$\Pi_A = \rho^{x_A} l^{y_A} U^{z_A} A \tag{3.2}$$

and

$$R = \frac{\rho U l}{\mu} \quad \text{(Reynolds number)} \tag{3.3}$$

while Γ is the ratio of two lengths

$$\Gamma = \frac{l_1}{l_2} \tag{3.4}$$

The ratio Γ reflects the influence of that part of the geometry of the system which can be varied (adjusted) and yet which can be satisfactorily defined by a single number (namely by the number Γ itself); the remaining 'undefinable part' of the geometry being reflected by the form φ_A of the function (3.1). Clearly, the functional relation (3.1) can be generalized to any number of dimensionless parameters of adjustable geometry

$$\Pi_A = \varphi_A(\text{Re}, \Gamma_1, \Gamma_2, \ldots) \tag{3.5}$$

Consider, for example, the airfoil moving close to a fixed plain surface with a constant velocity U directed parallel to the surface (Fig. 3.1). For a given geometry of the system, the mechanical structure of the flow is completely determined by the nature of the fluid (air), by the absolute

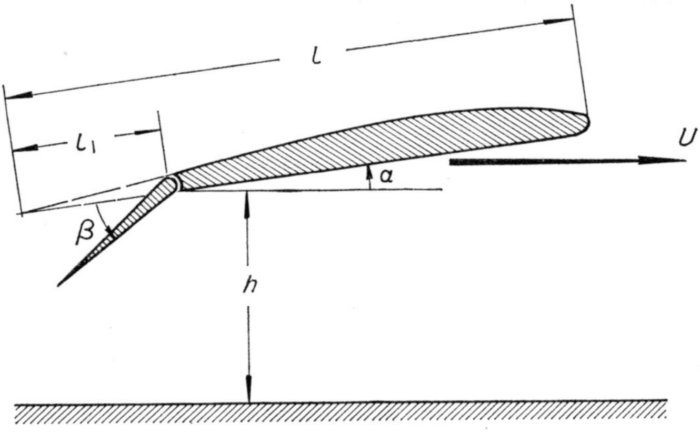

FIG. 3.1

size of the system and by the velocity U. Thus, the mechanical structure of the airflow is defined by the parameters

$$\mu, \rho, l, U$$

i.e. by the Reynolds number

$$R = \frac{Ul}{\nu} \qquad \left(\text{where } \nu = \frac{\mu}{\rho}\right)$$

On the other hand, the geometry of the system is determined by the shape of the airfoil (which cannot be defined by a limited number of parameters) and by the following components of definable (and adjustable) geometry

$$\begin{aligned} \Gamma_1 &= \alpha \text{ (angle of attack)} \\ \Gamma_2 &= h/l \text{ (relative elevation of the airfoil)} \\ \Gamma_3 &= l_1/l \text{ (relative length of the flap)*} \\ \Gamma_4 &= \beta \text{ (inclination angle of the flap)} \end{aligned} \tag{3.6}$$

Hence, for the airflow under consideration, the relation (3.5) becomes

$$\Pi_A = \varphi_A \left(\frac{Ul}{\nu}, \alpha, \frac{h}{l}, \frac{l_1}{l}, \beta\right) \tag{3.7}$$

* It is assumed that even though the shape of the airfoil is specified, the relative length l_1/l of its mobile part (flap) is under investigation.

where the form of the influence of undefinable shape of the airfoil is reflected by the form of the function φ_A. For example, if the property A under investigation is the drag force F_x or the lift force F_y, then (considering that the basic quantities are ρ, l and U) the eqn. (1.20) gives

$$\left.\begin{aligned}\Pi_x &= \rho^{-1}l^{-2}U^{-2}F_x = \frac{F_x}{\rho l^2 U^2}\\ &\text{and}\\ \Pi_y &= \rho^{-1}l^{-2}U^{-2}F_y = \frac{F_y}{\rho l^2 U^2}\end{aligned}\right\} \tag{3.8}$$

The dimensionless quantities Π_x and Π_y are referred to as *drag* and *lift coefficients* respectively. Thus, drag and lift coefficients must be certain functions of the Reynolds number and of the geometry of the system

$$\begin{aligned}\Pi_x &= \varphi_x\left(\frac{Ul}{\nu}, \alpha, \frac{h}{l}, \frac{l_1}{l}, \beta\right)\\ \Pi_y &= \varphi_y\left(\frac{Ul}{\nu}, \alpha, \frac{h}{l}, \frac{l_1}{l}, \beta\right)\end{aligned} \tag{3.9}$$

If the form of the functions φ_x and φ_y is revealed by the measurements carried out in a small scale model, then the magnitudes of the drag and the lift force acting on a full size wing (prototype), corresponding to the same geometry, can be obtained by multiplying the prototype value of $\rho l^2 U^2$ by φ_x and φ_y respectively. ← model ?

Let us consider, as a second example, the part of a rough closed conduit shown in Fig. 3.2 (which might possess a non-standard, say, elliptical, shape of the cross-section, and which might be considered for the design of a future installation). The geometry (texture) of its roughness might be very complicated, and thus one might not be able to define it by a limited number of parameters. However, the rest of the geometric properties of the system can be completely defined by the following dimensionless quantities.

$$\left.\begin{aligned}\Gamma_1 &= \frac{k}{l} \text{ (relative roughness)}\\ \Gamma_2 &= \frac{l_1}{l} \text{ (aspect ratio of elliptic cross-section)}\\ \Gamma_3 &= \frac{r}{l} \text{ (relative curvature)}\\ \Gamma_4 &= \alpha \text{ (angle of the bend)}\\ \Gamma_5 &= \beta \text{ (angle of reduction)}\end{aligned}\right\} \tag{3.10}$$

Thus, the dimensionless version Π_A corresponding to any property A of the flow in the closed conduit shown in Fig. 3.2 must be given by the functional relation

$$\Pi_A = \varphi_A\left(\frac{Ul}{\nu}, \frac{k}{l}, \frac{l_1}{l}, \frac{r}{l}, \alpha, \beta\right) \tag{3.11}$$

where the typical velocity U can be, for example, the average velocity of, say, the larger straight part of the conduit. If the property A under in-

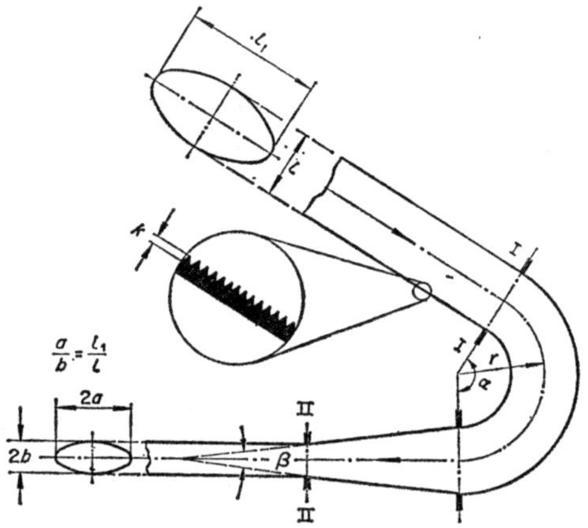

FIG. 3.2

vestigation is the pressure drop between the sections I-I and II-II, then applying (1.20) we obtain

$$\Pi_A = \rho^{-1}l^0U^{-2}(p_{\mathrm{I}} - p_{\mathrm{II}}) = \frac{p_{\mathrm{I}} - p_{\mathrm{II}}}{\rho U^2} \tag{3.12}$$

Substituting this value of Π_A in (3.11) we arrive at

$$\frac{p_{\mathrm{I}} - p_{\mathrm{II}}}{\rho U^2} = \varphi_p\left(\frac{Ul}{\nu}, \frac{k}{l}, \frac{l_1}{l}, \frac{r}{l}, \alpha, \beta\right) \tag{3.13}$$

i.e.

$$\frac{p_{\mathrm{I}} - p_{\mathrm{II}}}{\gamma} = \zeta\frac{U^2}{2g} \tag{3.14}$$

where $\gamma = \rho g$ is the specific weight of the fluid, while ζ implies $2\varphi_p$. Hence, the coefficient ζ of the pressure energy loss in the closed conduit shown in Fig. 3.2 must be a function of the Reynolds number and of five variables Γ_j defining the geometry of the system; the influence of the undefinable geometry (texture) of roughness being reflected by the form of the function φ_p.

Clearly, the dynamic similarity of the phenomena such as those described here, i.e. which are determined by the Reynolds number Re and by the geometry of the system (the form of φ_A and Γ_j), is given by the geometric similarity and by the identity of model and prototype Reynolds numbers, i.e. by

$$\lambda_{\mathrm{Re}} = 1 \tag{3.15}$$

A geometrically similar model which satisfies the condition (3.15) will be referred to henceforth as a *Reynolds model.* All the experimental curves representing the variation of any dimensionless property Π_A with any of the dimensionless variables Re and Γ_j determined by the measurements in a Reynolds model can equally well be used for the prototype also. The prototype value A' of any dimensional property of the phenomenon can be obtained either by multiplying the measured model value A'' with the scale of this property λ_A as

$$A' = \lambda_A A'' \tag{3.16}$$

or, by multiplying the determined dimensionless property Π_A by the corresponding power product (formed by the known prototype characteristic parameters)

$$\rho'^{x_A} l'^{y_A} U'^{z_A}$$

as

$$A' = (\rho'^{x_A} l'^{y_A} U'^{z_A})\Pi_A \tag{3.17}$$

We will now proceed to consider the determination of the scales λ_A which must be known when converting the values according to (3.16).

3.2 Scale Relations for Reynolds Models

Considering the value of the Reynolds number (3.3), the criterion of dynamic similarity (3.15) can be expressed as follows:

$$\lambda_\rho \lambda_U \lambda_l \lambda_\mu^{-1} = 1 \tag{3.18}$$

Similarly, using the value of Π_A given by (3.2) we can express the condition

$$\lambda_{\Pi_A} = 1$$

in the following (open) form

$$\lambda_\rho^{x_A} \lambda_U^{y_A} \lambda_l^{z_A} \lambda_A = 1 \qquad \text{(3.19}$$

Theoretically, the scales of any three quantities having independent dimensions can be selected freely. For example, the quantities μ, ρ and l possess independent dimensions, and thus the scales λ_μ, λ_ρ and λ_l can be chosen freely. In this case, the Reynolds criterion (3.18) gives the following value for the velocity scale

$$\lambda_U = \lambda_\mu \, \lambda_\rho^{-1} \, \lambda_l^{-1} \tag{3.20}$$

Eliminating λ_U between (3.20) and (3.19) we arrive at the expression of the scale λ_A in terms of the selected scales λ_μ, λ_ρ and λ_l

$$\lambda_A = \lambda_\mu^{-y_A} \, \lambda_\rho^{(y_A - x_A)} \, \lambda_l^{(y_A - z_A)} \tag{3.21}$$

In practice, however, the model usually operates with the prototype fluid. In other words, in practice we 'choose' the scales of μ and ρ as

$$\lambda_\mu = 1 \quad \text{and} \quad \lambda_\rho = 1 \tag{3.22}$$

Accordingly (3.20) reduces into

$$\lambda_U = \frac{1}{\lambda_l} \tag{3.23}$$

while (3.21) becomes

$$\lambda_A = \lambda_l^{(y_A - z_A)} \tag{3.24}$$

where the remaining third scale λ_l can still be selected freely.

Since dynamically similar models must necessarily be geometrically similar (or since the dynamic similarity is given by the identity of all the dimensionless variables) it is obvious that

$$\lambda_{\Gamma_j} = 1 \qquad (j = 1, 2, \ldots) \tag{3.25}$$

must also be valid in addition to $\lambda_{Re} = 1$. From (3.20) or (3.23), it is clear that λ_U is inversely proportional to λ_l, i.e. the *smaller* the model the *higher* are the model velocities. This is characteristic of the Reynolds models.

3.3 Roughness

Let one of the dimensionless parameters Γ_j be the relative roughness (as in the second example concerning the flow in the closed conduit), e.g. let us assume that

$$\Gamma_1 = \frac{k}{l} \tag{3.26}$$

According to (3.25) we must have

$$\lambda_{\Gamma_1} = 1$$

i.e.

$$\lambda_k = \lambda_l \tag{3.27}$$

In other words, the concept of geometric similarity of a Reynolds model, strictly speaking, includes the geometric similarity of roughness. Technically, it is very difficult to provide the model with geometrically similar roughness, even if such a similarity is implied in the statistical sense. Accordingly, in practice one attempts to avoid this difficulty whenever possible. The purpose of the present paragraph is to reveal those cases where it is permissible to substitute the geometrically similar roughness by a different, and yet with respect to its influence, equivalent roughness.

Let us assume, for example, that the flow under investigation is always in laminar regime. As is well known, if the roughness size (height) k is small in comparison to the smallest dimension l of the flow (i.e. if the ratio $\Gamma_1 = k/l$ can be regarded as infinitesimal of the first order), then the velocity distribution (and thus the mechanical structure) of the laminar flow does not depend on $\Gamma_1 = k/l$. Accordingly, the ratio $\Gamma_1 = k/l$ is no longer the dimensionless variable of the phenomenon, and thus the condition (3.27) can be relaxed. In such cases, the experiments can be carried out with any model roughness that can be regarded as small in comparison to the smallest dimension of the model flow. In fact, they can be carried out for the extreme case $k = 0$, that is, in practically smooth models made of glass, plexiglass, etc.

The situation becomes considerably more difficult if the flow under investigation is (or can become) turbulent. Before studying the problem of roughness for models operating with turbulent flow, it may be well to refresh our memory on some related concepts.

As is well known, the distribution of the velocities u in the vicinity of the flow boundaries is given by the following logarithmic formula ('wall law' of L. Prandtl)

$$\frac{u}{v_*} = \frac{1}{\kappa} \ln \frac{y}{k} + B \tag{3.28}$$

where $\kappa \approx 0{\cdot}4$ is the Von Karman constant,* $v_* = \sqrt{(\tau_0/\rho)}$ the shear velocity, y the distance from the flow boundary, k the height of the

* $\kappa = 0{\cdot}4$ is valid for any homogeneous fluid and for any geometry of roughness (provided that $\Gamma_1 = k/l$ is small). Hence, the term 'universal constant' (of Von Karman). If, however, the fluid is not homogeneous, then, it follows from the experimental measurements, that the value of Von Karman 'constant' can vary considerably[1]

protrusions forming the roughness and B is a function of the Reynolds number $v_* k/\nu$ which can be denoted as follows

$$B = \varphi_B \left(\frac{v_* k}{\nu}\right) \tag{3.29}$$

The experimental curve representing the function (3.29) for the sand roughness k_s (or to be more precise, for any roughness possessing the geometry that can be regarded as statistically identical to that of the sand roughness) is shown in Fig. 3.3. The following three zones are typical for

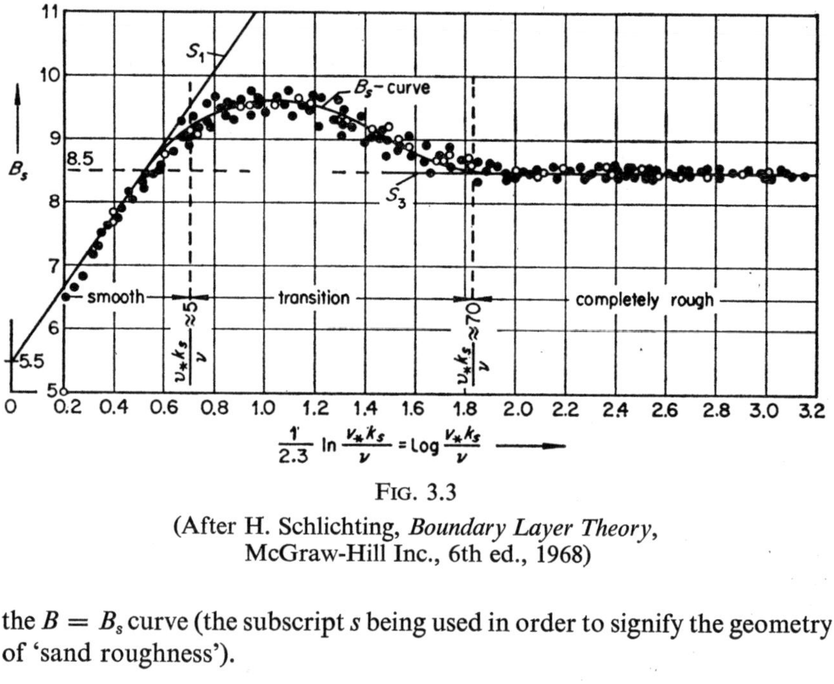

FIG. 3.3

(After H. Schlichting, *Boundary Layer Theory*, McGraw-Hill Inc., 6th ed., 1968)

the $B = B_s$ curve (the subscript s being used in order to signify the geometry of 'sand roughness').

Zone I ($v_* k_s/\nu < 5$) where the curve B_s is indistinguishable from the straight line S_1 given by

$$(B_s)_1 = \frac{1}{\kappa} \ln \frac{v_* k_s}{\nu} + 5{\cdot}5 \tag{3.30}$$

This zone corresponds to the 'hydraulically smooth' regime of the turbulent flow, which does not depend on roughness. Observe that if (3.30) is substituted in (3.28), then $k = k_s$ vanishes from the expression of the velocity u.

Zone II ($\approx 5 < v_* k_s/\nu < \approx 70$) where the curve B_s cannot be given by a simple analytical form, and thus where the value of B_s must be estimated from Fig. 3.3 depending on the value of the Reynolds number $v_* k_s/\nu$. This zone corresponds to the transitional regime (of the turbulent flow) which is dependent on both viscosity and roughness. Observe that if the general form (3.29), which can be regarded as $(B_s)_2$, is substituted in (3.28), then both ν and $k = k_s$ will be present in the expression of u.

Zone III ($v_* k_s/\nu > \approx 70$) where the curve B_s is indistinguishable from the straight line S_3 implying

$$(B_s)_3 = 8{\cdot}5 \tag{3.31}$$

This zone corresponds to the 'fully developed' regime of turbulent flow, or simply to 'rough turbulent flow', which does not depend on viscosity. (Substitute (3.31) in (3.28) and note that μ is not present in the expression of u.)

If the geometry of a roughness k is different from the geometry of the sand roughness k_s, then the curve B corresponding to the roughness k will also be different from the curve B_s corresponding to k_s. However, the general trend (the character) of all B curves will remain essentially the same. Indeed, all B curves must, for example, merge into the same straight line S_1 for small values of $v_* k/\nu$, because whatever the geometry of the (small in comparison to l) roughness might be, the decrement of the Reynolds number $v_* k/\nu$, and thus the increment of the relative thickness $\bar{\delta}/k$ of the viscous sublayer, must eliminate the influence of k on u. On the other hand, the parameter k can disappear from the expression (3.28) only if the function B possesses the form (3.30) implying the straight line S_1. Similarly, whatever the geometry of the roughness might be, when the values of $v_* k/\nu$ are sufficiently large, then the influence of the viscosity ν, and thus of the Reynolds number $v_* k/\nu$, must vanish, and thus the B curves must become parallel to the abscissa $v_* k/\nu$. Hence, two different curves B_a and B_b corresponding to two different types of roughness geometry a and b must exhibit the trends shown schematically in Fig. 3.4. Clearly, there is no reason to expect that the curves B_a and B_b should merge into the common straight line S_1 and into 'their own' straight lines $(S_3)_a$ and $(S_3)_b$ for the same values of $v_* k/\nu$. Thus, the limits of the transitional regime for a and b are deliberately indicated in Fig. 3.4 by different points C_a, C_b and D_a, D_b. It is thus clear that the numerical values ≈ 5, ≈ 70 and 8·50 are valid only for sand roughness (or for the roughness having the same statistical geometry as sand roughness). On the other hand, the numerical value 5·5 which appears in the relation (3.30) is valid for any roughness.

Now we can consider the question of whether it is possible to reproduce the prototype phenomenon having the geometry of roughness that can be denoted by a in a model having a different roughness geometry, say, b. From boundary layer theory it is known that, the fluid motion sufficiently far from the flow boundaries is indistinguishable from the frinctionless potential flow, which does not depend on the roughness of the flow boundaries. Hence, any problem related to the influence of roughness must be

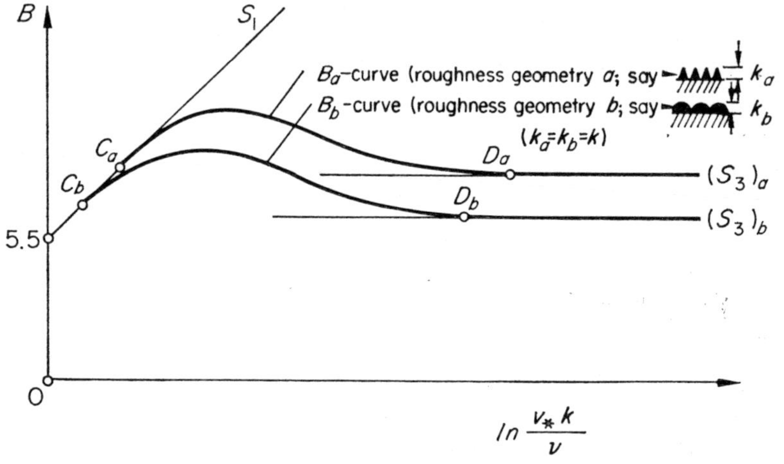

FIG. 3.4

considered with respect to the flow in the vicinity of the boundaries. The velocity distribution in the prototype flow near the boundaries is given by

$$\frac{u'}{v_*'} = \frac{1}{\kappa} \ln \frac{y'}{k_a'} + B_a' \tag{3.32}$$

the same in the model flow being given by

$$\frac{u''}{v_*''} = \frac{1}{\kappa} \ln \frac{y''}{k_b''} + B_b'' \tag{3.33}$$

The dynamic similarity of the model and the prototype requires that the dimensionless velocities u'/v_*' and u''/v_*'' are identical in all the corresponding points (given in the corresponding sections by the identical values of the dimensionless distances y'/l' and y''/l''). Equating (3.32) and (3.33) we obtain

$$\frac{1}{\kappa} \ln \frac{y'}{k_a'} + B_a' = \frac{1}{\kappa} \ln \frac{y''}{k_b''} + B_b''$$

or

$$\frac{1}{\kappa} \ln \frac{y'}{l'} + \frac{1}{\kappa} \ln \frac{l'}{k_a'} + B_a' = \frac{1}{\kappa} \ln \frac{y''}{l''} + \frac{1}{\kappa} \ln \frac{l''}{k_b''} + B_b''$$

and since this equality is supposed to be valid for identical values of y'/l' and y''/l'' we arrive at

$$\frac{1}{\kappa}\ln\frac{l'}{k_a'} + B_a' = \frac{1}{\kappa}\ln\frac{l''}{k_b''} + B_b'' \tag{3.34}$$

and thus

$$\frac{k_b''}{k_a'} = \frac{l''}{l'} \times e^{-\kappa(B_a' - B_b'')} \tag{3.35}$$

Here, the ratio l''/l' is the geometric model scale whereas k_b''/k_a' is the roughness height scale λ_k, i.e. (3.35) can be expressed as follows

$$\lambda_k = \lambda_l \times e^{-\kappa(B_a' - B_b'')} \tag{3.36}$$

Hence we have arrived at a scale relation that is of a more general nature than (3.27). Indeed, the eqn. (3.36) reduces into (3.27) only if

$$B_a' \equiv B_b''$$

is valid.

Clearly, the scale relation (3.36) can have a practical meaning only if the roughness scale λ_k has a certain constant value. This, however, becomes possible only if the difference $(B_a' - B_b'')$ has a certain constant value. From Fig. 3.4 it follows clearly that the difference between the B values can remain constant only if neither model nor prototype flows are in the transitional regime.

(i) If the model and prototype flows are both in the hydraulically smooth regime, then

$$B_a' - B_b'' = (B_1')_a - (B_1'')_b = 0 \tag{3.37}$$

and thus

$$\lambda_k \equiv \lambda_l \tag{3.38}$$

will always be valid, provided the model and prototype values of v_*k/ν are identical (which will necessarily be so since $\lambda_{Re} \equiv 1$).* Observe that (3.38) is valid independently of the geometry of the model and prototype roughness, for the right hand side of (3.38) does not contain any quantity that depends on the roughness geometry. Hence, in the case of the hydraulically smooth regime of turbulent flow, the roughness is as unimportant as in the case of laminar flow.

* Put $\lambda_l = \lambda_k$ and $\lambda_U = \lambda v_*$ in $\lambda_{Re} = \lambda_l\lambda_U\lambda_\rho\lambda_\mu^{-1} = 1$ and observe that the relation obtained implies that: the scale of the Reynolds number v_*k/ν is equal to unity.

(ii) If the model and prototype flows are both in rough turbulent regimes, then we have

$$B_a' - B_b'' = (B_3')_a - (B_3'')_b = \text{const} \gtrless 0 \tag{3.39}$$

and thus

$$\lambda_k = \lambda_l \times e^{\text{const}.\kappa} \tag{3.40}$$

even if the model and prototype values of $v_* k/v$ are not necessarily identical. In (3.40) the value of the constant might be positive or negative, it depends entirely on the roughness geometry of the model and prototype. Accordingly, λ_k might be larger or smaller than λ_l. In the conventional model technique, however, usually

$$\lambda_k > \lambda_l \tag{3.41}$$

which implies that $(B_3')_a < (B_3'')_b$.

The relation (3.36), which was so useful in the explanations above, unfortunately cannot be used for the design of practical models, for there are hardly any B-values available for roughness geometry other than that of sand roughness. Therefore, in a case of rough turbulent flow, the model roughness is almost invariably determined by an experimental trial and error method which can be explained as follows. Let us assume that, for a certain stage of the prototype flow, the value of the typical velocity U' and the values of the velocities u_i', corresponding to some prototype points m_i' are known (either from prototype measurements or from theory). Clearly, if the model roughness is incorrect, then by setting the model velocity

$$U'' = \lambda_U U' \tag{3.42}$$

we shall not be able to have, at the corresponding model points m_i'', the required values

$$u_i'' = \lambda_U u_i' \tag{3.43}$$

(for any i). Thus, the roughness of the model surface is adjusted until the required equalities (3.43) are achieved.

In the transitional regime, as has been pointed out earlier, the difference $(B_a' - B_b'')$ varies with the stage of the flow (with the Reynolds number $v_* k/v$), and thus every different stage of the flow requires its own value of the scale λ_k, and thus of k_b''. Accordingly, the experimental determination of the model roughness described above for rough turbulent regime can be adopted for transitional regime, strictly speaking, only for a single stage of the flow (for a single value of $v_* k/v$). If the investigation of a wide range of transitional flows is required, one has no alternative but to use either the geometrically similar roughness ($\lambda_k = \lambda_l$) or subdivide the wide range of $v_* k/v$ into a number of smaller ranges and investigate each of them separately by using their own different k_b'' (or their own different configurations of the same k_b'', especially if the roughness is formed by the identical elements glued on the surface).

If the values of the *equivalent sand roughness* corresponding to the rough surfaces of the model and prototype are known, then in cases of rough turbulent flow, the model roughness can be determined without recourse to the experimental trial and error method. One determines (from a table such as Table 3.1) the model and prototype values k_s'' and k_s' of the equivalent sand roughness so that their ratio is equal to the model scale λ_l; i.e. as

$$\lambda_l = \frac{k_s''}{k_s'} = \lambda_k \tag{3.44}$$

Indeed, since the roughness in both model and prototype is now represented by the same geometry (of the sand roughness), we have

$$(B_3')_a = (B_3'')_b = 8{\cdot}5$$

Accordingly, the relation (3.39) gives const = 0, while (3.40) gives $\lambda_k = \lambda_l$.

Table 3.1

Nature of flow boundary	Equivalent sand roughness k_s (ft)
Copper, lead, brass, alkathene, glass, asbestos cement	<0·00005 (smooth)
Cast iron, bitumen lined	0·0001
Cast iron, concrete lined	0·0001
Uncoated steel	0·0001
Coated steel	0·0002
Galvanised iron	0·0005
Coated cast iron	0·0005
Uncoated cast iron	0·001
Wet-mix spun precast concrete	0·002
Glazed stoneware	0·002
Precast concrete, mortar not wiped on inside of joint	0·01

After N. Webber, *Fluid Mechanics for Civil Engineers*, E. & F. N. Spon, 1968)

3.4 Large Values of the Reynolds Number

If the flow is rough turbulent, then the viscosity μ is no longer a characteristic parameter, while the Reynolds number $R = \rho Ul/\mu$ (which reflects the influence of μ) is no longer a dimensionless variable. Hence, if the flow is rough turbulent, then the relation (3.5) reduces into

$$\Pi_A = \varphi_A(\Gamma_1, \Gamma_2, \ldots) \tag{3.45}$$

which implies that the dynamic similarity is determined solely by the *geometric similarity* (i.e. by the identity of the form of the function and by equality of the geometric variables Γ_j). The geometric similarity mentioned includes the similarity of roughness, for, in accordance with (3.26), it is assumed that Γ_1 represents the relative roughness.

From the preceding considerations, it is clear that if the flow is rough turbulent, then, whatever the geometry of the given roughness, the consideration of the given roughness (having the height k) can always be replaced by that of the equivalent sand roughness (having the height k_s). The relations that follow are expressed in terms of k_s, regardless of whether the actual roughness of the flow boundaries is (or can be regarded as) the sand roughness, or it is different from the sand roughness and k_s represents the equivalent value. Accordingly, Γ_1 implies

$$\Gamma_1 = \frac{k_s}{l} \tag{3.46}$$

while (3.45)

$$\Pi_A = \varphi_A\left(\frac{k_s}{l}, \Gamma_2, \ldots\right) \tag{3.47}$$

The existence of rough turbulent flow means that

$$\frac{v_* k_s}{\nu} > \approx 70 \tag{3.48}$$

The Reynolds number $v_* k_s/\nu$ in the inequality above is related to the Reynolds number Re by the following equation

$$\mathrm{Re} = \frac{Ul}{\nu} = \frac{U}{v_*} \cdot \frac{l}{k_s} \cdot \frac{v_* k_s}{\nu} \tag{3.49}$$

Here, the ratio U/v_* is the friction factor c, which in the case of a rough turbulent flow, is an increasing function of l/k_s alone. Let us denote this function by

$$c = \frac{U}{v^*} = \varphi_c\left(\frac{l}{k_s}\right) \tag{3.50}$$

In this case the eqn. (3.49) can be expressed as

$$\mathrm{Re} = \varphi\left(\frac{l}{k_s}\right)\frac{v_* k_s}{\nu} \tag{3.51}$$

where

$$\varphi\left(\frac{l}{k_s}\right) = \frac{l}{k_s}\varphi_c\left(\frac{l}{k_s}\right) \tag{3.52}$$

Eliminating $v_* k_s/\nu$ from (3.51) and (3.48), we arrive at the following condition of existence of a rough turbulent flow, given in terms of the Reynolds number Re

$$\mathrm{Re} > \approx 70\,\varphi\left(\frac{l}{k_s}\right) \tag{3.53}$$

Hence, it is not possible to associate the initiation of a rough turbulent flow with a single value of the Reynolds number Re. The value of Re corresponding to the initiation of rough turbulent flow depends on the relative roughness k_s/l. From the fact that $\varphi(l/k_s)$ is an increasing function of its variable l/k_c (relative smoothness), it follows that: the smoother the flow boundaries, the larger the value of Re corresponding to the beginning of rough turbulent flow.

It is clear that, in order to ensure the presence of rough turbulent flow in all parts of the system, the inequality (3.48) must be satisfied even by the smallest value of the shear velocity $v_* = \sqrt{(\tau_0/\rho)}$. In other words, if the value of the shear stress τ_0 acting on the flow boundary varies from one point of the flow boundary to another, then v_*, which appears in (3.48), must be formed by the smallest value of τ_0. Obviously, for given μ, ρ, U and the size l of the system, the value and location of the smallest τ_0, and thus v_*, depend entirely on the geometry of the system. Hence, the value of the ratio U/v_*, that is the value of the function φ_c, and consequently of φ, is also determined by the geometry of the system. Denoting the value of $70 \times \varphi$ by $\mathrm{Re}_{\min}$ we can express the condition (3.53) as

$$\mathrm{Re} > (\mathrm{Re})_{\min} \tag{3.54}$$

where the value of $(\mathrm{Re})_{\min}$, which corresponds to the initiation of rough turbulent flow in the system, is itself a function of the geometry of the system (relative roughness being covered by the term 'geometry').

It follows that if the flow in both model and prototype is rough turbulent and thus if Re is no longer the variable in either of these flow systems, then the condition

$$\lambda_{\mathrm{Re}} = 1$$

can be relaxed, that is, the investigation can be carried out in a geometrically similar model even if Re″ and Re′ are not equal. All that is required is that both Re″ and Re′ are larger than a certain minimum Re_{min} (determined by the geometry of the system), i.e. that the condition

$$(Re)_{min} < Re'' < Re' \tag{3.55}$$

is satisfied. The inequality $Re'' \neq Re'$ can only be helpful with regard to model tests if it gives for the model velocities u'' a value that is smaller than $u'' = u'(l'/l'')$ given by the usual Reynolds models ($Re'' = Re'$). Hence it is assumed automatically that the inequality $Re'' \neq Re'$ provides $u'' < u'(l'/l'')$ and thus $l''u'' < l'u'$ which, for a model operating with the prototype fluid, is equivalent to $Re'' < Re'$ as adopted in (3.55). If the system possesses a simple geometry, then the determination of $(Re)_{min}$ is also simple. Consider, for example, the flow in a straight circular pipe. Let us assume that typical length l is the pipe diameter D, while the typical velocity U is the average velocity v. The value of τ_0, and thus of v_* is the same at any point of the flow boundary, while the ratio v/v_* is given by the following well-known relation

$$\frac{v}{v_*} = \varphi_c \left(\frac{D}{k_s}\right) = \frac{1}{\kappa} \ln \left(7{\cdot}8 \frac{D}{k_s}\right) \tag{3.56}$$

Thus

$$\varphi \left(\frac{D}{k_s}\right) = \frac{1}{\kappa} \frac{D}{k_s} \ln \left(7{\cdot}8 \frac{D}{k_s}\right)$$

which gives at once

$$(Re)_{min} = 70\, \varphi \left(\frac{D}{k_s}\right) = 175 \frac{D}{k_s} \ln \left(7{\cdot}8 \frac{D}{k_s}\right) \tag{3.57}$$

For example, if $D = 10$ cm, and $k_s = 1$ mm and the equation above gives

$$(Re)_{min} = 175 \times 100 \times \ln (780) = 166\,000$$

If the fluid is water (in 15°C) then $\nu \approx 1{\cdot}01 \times 10^{-6}$ m²/s and we obtain

$$v_{min} = (Re)_{min} \times \frac{\nu}{D} = 166\,000 \times \frac{1{\cdot}01 \times 10^{-6}}{0{\cdot}1} = 1{\cdot}8 \text{ m/s}$$

Hence, any flow in the circular pipe considered above that has an average velocity not less than 1·8 m/s will belong to the range

$$Re > (Re)_{min}$$

However, the geometry of the systems encountered in the practice is usually far more complicated, and one has no alternative but to determine the values of $(Re)_{min}$ experimentally. The simplest and most reliable way is to measure the magnitudes U_1'', U''_2, . . . U_N'' of the typical velocity U'' corresponding to the sequence 1, 2, . . . N of increasing (with regard to their intensity) stages of the flow, and at the same time to measure the magnitudes u_1'', u_2'', . . . u_N'' of a velocity u'' at a fixed point of the space occupied by model flow. (Clearly, the fixed point mentioned must be selected in the slowest part of the flow. A more reliable approach would be to repeat the procedure for a series of the points selected in various parts of the flow.) This point should be selected as far as possible from the location of the typical velocity and as near as possible to the object undergoing the model test. From the measured values, we can form the following sequence of dimensionless velocities

$$\frac{u_1''}{U_1''};\ \frac{u_2''}{U_2''};\ \ldots\ \frac{u_N''}{U_N''} \tag{3.58}$$

After a certain index n $(1 < n < N)$ the dimensionless velocities above become independent of the Reynolds number, i.e. they become equal. The stage signified by the index n is that which corresponds to the Reynolds number

$$(\mathrm{Re})_{\mathrm{min}} = \frac{Un''l''}{\nu} \tag{3.59}$$

It is thus clear that the above method of determining $(\mathrm{Re})_{\mathrm{min}}$ rests on revealing the stage which forms the boundary between the varying and non-varying with the stage (i.e. with the Reynolds number) ratios (3.58).

Some relevant points concerning the Reynolds models

(i) Observe (from (3.52) and (3.57)) that when k_s/l approaches zero, then the function $\varphi(l/k_s)$ tends to infinity, i.e. that

$$\lim_{k_s/l \to 0} \varphi\left(\frac{l}{k_s}\right) \to \infty \tag{3.60}$$

is valid (at least because of the presence of the multiplier l/k_s in the expression of $\varphi(l/k_s)$). This property can also be seen, for example, from (3.57). Thus, if the flow boundaries are practically smooth (glass, plexiglas, etc.), then the value of $(\mathrm{Re})_{\mathrm{min}}$ is practically infinite. It follows that, the method described in the present section cannot be applied for flows past smooth surfaces. For example, from the family of curves in Fig. 1.4 one can see clearly that the smoother the pipe, the larger the Reynolds number Re, beyond which the λ-curve degenerates into a straight line parallel to the abscissa Re (beyond which λ becomes independent of Re): the λ-curve corresponding to the smooth pipe never being parallel to the abscissa Re.

(ii) The condition (3.55), which indicates the region of application for the method described in this section, is valid as it stands if separation of the flow from the boundaries is impossible (parallel flows, converging flows, flows past slender rough bodies etc.). If separation of the flow is possible (flow past blunt bodies), then a region which can be regarded as independent of Re may be present even if the body is smooth and tests with non-equal values of Re can still be performed. However, in such

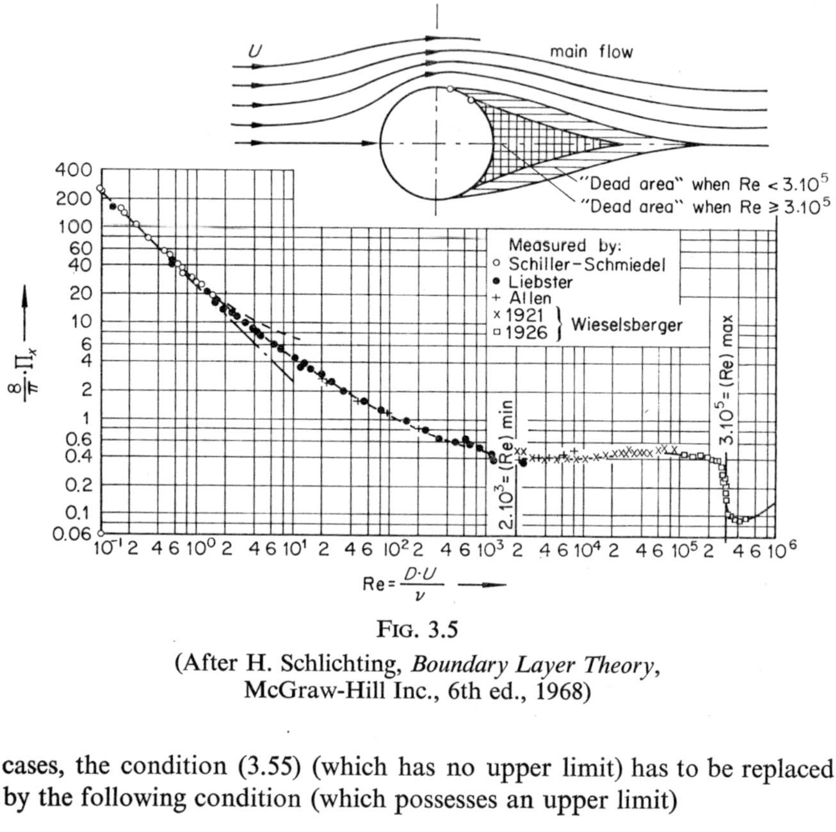

FIG. 3.5

(After H. Schlichting, *Boundary Layer Theory*, McGraw-Hill Inc., 6th ed., 1968)

cases, the condition (3.55) (which has no upper limit) has to be replaced by the following condition (which possesses an upper limit)

$$(\mathrm{Re})_{\min} < \mathrm{Re}'' < \mathrm{Re}' < (\mathrm{Re})_{\max} \qquad (3.61)$$

Consider, for example, the curve representing the variation of the drag coefficient

$$\Pi_x = F_x/\rho D^2 U^2$$

with the Reynolds number $\mathrm{Re} = UD/\nu$ for the sphere (blunt body) shown in Fig. 3.5. The curve in Fig. 3.5 indicates, that the value of Π_x can be regarded as independent of Re if it is larger than ≈ 2000, but smaller than $\approx 300\,000$. Hence for the flow past a sphere the condition (3.61) which must be satisfied by (not necessarily equal) model and prototype Reynolds numbers can be expressed as follows:

$$2000 < \mathrm{Re}'' < \mathrm{Re}' < 300\,000 \tag{3.62}$$

The validity of (3.55) rests on the presence of a rough turbulent flow, and thus on the independence of viscosity. The fact that for a certain range of Reynolds numbers the flow past blunt bodies can be regarded as being independent of Re is not because the flow past the body is rough turbulent in that range. Indeed, in the case of blunt bodies, firstly the roughness is of no relevance; secondly, it is generally accepted that the flow in the boundary layer surrounding a blunt body becomes turbulent only after the value of Re has exceeded the order $(\mathrm{Re})_{\max}$ (where the drag curve suddenly begins to drop down). In other words, the validity of (3.61) *ends* when turbulence *begins*. The sudden drop of the drag curve (referred to as 'drag crisis') has the following explanation. When the Reynolds number Re reaches a certain large value, namely $(\mathrm{Re})_{\max}$, then the boundary layer flow becomes turbulent and the configuration of eddies in the wake of separation (dead area) suddenly changes. The presence of turbulence implies the reduction of the influence of μ. Accordingly, the eddy configuration changes so as to bring the conditions of the flow nearer to those of the ideal fluid ($\mu = 0$), that is, so as to reduce the size of the wake of separation and thus the magnitude of the drag force (Fig. 3.5). Hence the sudden drop of the drag curve.*

Flows where no separation can be present, and so can be treated in accordance with (3.55) are distinguished by the fact that their pressure p always decreases in the direction x of the flow, i.e.

$$\frac{\partial p}{\partial x} < 0 \tag{3.63}$$

is valid for any part of the space occupied by the flow (flows through straight conduits, converging wedges, etc.). If, in some parts of the space occupied by the flow the pressure gradient is positive

$$\frac{\partial p}{\partial x} > 0 \tag{3.64}$$

then the occurrence of separation is possible (flows past blunt obstacles, diverging conduits, etc.).** Such flows can be treated only in accordance

* An extensive account on this subject which includes the first explanation of L. Prandtl[5], as well as the most recent version, derived from the measurements carried out by A. Roshko[6] can be found e.g. in the works of G. K. Batchelor[4] and H. Schlichting[2].

** For blunt bodies 'friction drag' due to roughness is negligible in comparison to 'form drag' (due to geometry). See e.g. Ref. 4.

with (3.52). The numerical value of $(\mathrm{Re})_{max}$ corresponding to a given system can be determined by an experimental procedure completely analogous to the determination of $(\mathrm{Re})_{min}$ (eqns. (3.58), (3.59)).

(iii) Sometimes the system under investigation cannot be treated as a single geometrical totality which can be adequately represented by a single typical length l, and thus by a single Reynolds number formed by such a length. Consider, for example, the case of an aircraft, which would

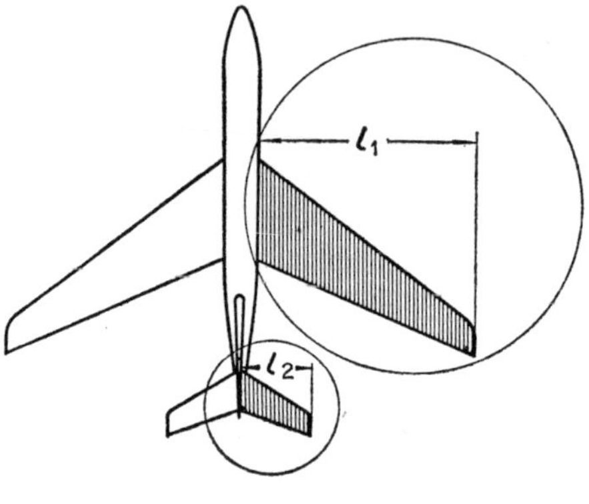

FIG. 3.6

be a typical subject for the investigations carried out in a Reynolds model for large values of Re. How should one form the Reynolds number for an aircraft? If the attention is focused on the wings, then it would be reasonable to take as the typical length l_1 (Fig. 3.6) and form the Reynolds number as

$$\frac{Ul_1}{\nu} \tag{3.65}$$

But how much sense can be derived from a graph where the measured values of, say, drag force acting on the tail plane are plotted versus Ul_1/ν (where l_1 is, for example, three times larger than the typical length l_2 of the tail plane)? Even though the geometry of the aircraft is specified, and it is thus known that l_1 and l_2 are related by a certain constant proportion, it is still considerably more reasonable to plot the values of the drag force acting on the wings and on the tail plane versus their own Reynolds numbers

$$\frac{Ul_1}{\nu} \quad \text{and} \quad \frac{Ul_2}{\nu} \tag{3.66}$$

From the above example it follows that sometimes it might be advisable to divide the system into certain sub-systems and to investigate them separately to a certain extent.

3.5 High Speed Reynolds Models

It might be worth while to mention briefly some properties of the Reynolds models operating at high speed, even if such models find their application in the field of aircraft engineering rather than in hydraulic engineering. When the flow velocity U becomes comparable with the velocity of sound C (in that fluid), then the progress of the flow phenomenon becomes dependent on the additional dimensionless variable

$$\mathrm{Ma} = \rho^0 l^0 U^{-1} C = \frac{C}{U} \tag{3.67}$$

called *Mach number*. The beginning of the influence of the Mach number varies, depending on the value of Mach number itself, on the nature of the property A under investigation, and on the geometry of the system. To give an idea on the values of Mach number beyond which its influence becomes noticeable, the value Ma $\approx 0{\cdot}3$ can be mentioned. If the occurrence of separation is possible, then for Re $>$ (Re)$_{\max}$ and, say, Ma $> \approx 0{\cdot}3$ we have, in general, the influence of both Reynolds and Mach numbers and the expression of a dimensionless property becomes

$$\Pi_A = \varphi_A(\mathrm{Re}, \mathrm{Ma}, \Gamma_1, \Gamma_2, \ldots) \tag{3.68}$$

Thus, in a geometrically similar model (including roughness), the following Reynolds and Mach criteria of similarity must be satisfied

$$\lambda_{\mathrm{Re}} = 1;\ \lambda_{\mathrm{Ma}} = 1 \tag{3.69}$$

which give (see eqn. (3.20)

$$\lambda_U = \lambda_\mu \lambda_\rho^{-1} \lambda_l^{-1} \quad \text{and} \quad \lambda_U = \lambda_C \tag{3.70}$$

If the model is operating with the fluid of the prototype (air), then

$$\lambda_\mu = \lambda_\rho = \lambda_C = 1 \tag{3.71}$$

and the relations (3.69) become

$$\lambda_U = \frac{1}{\lambda_l} \quad \text{and} \quad \lambda_U = 1 \tag{3.72}$$

which imply that

$$\lambda_l = 1 \tag{3.73}$$

i.e. the model must be the size of the prototype.

Both $\lambda_{Re} = 1$ and $\lambda_{Ma} = 1$ can be maintained and yet $\lambda_U = \lambda_c > 1$, and thus $\lambda_l < 1$ (small scale model) can be achieved by using a different gas (e.g. Freon) in the model. However, this method of reduction of model size (and thus of wind tunnel costs) has not found a general acceptance. The reason for this lies in the fact that the thermodynamic properties, which are also of great importance in the design of high speed aircraft, are different for different gases and thus even if the model can be regarded as acceptable with respect to fluid mechanics, it may not be so thermodynamically.

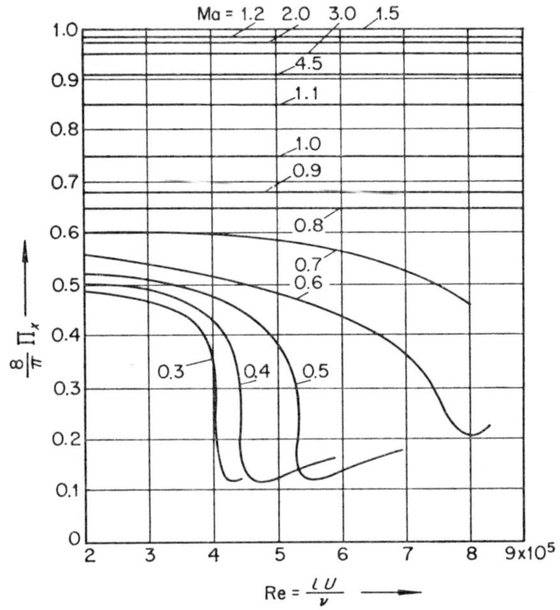

FIG. 3.7

(After H. Schlichting, *Boundary Layer Theory*, McGraw-Hill Inc., 6th ed., 1968)

In practice, one attempts to achieve a small scale model by accepting that the Reynolds and Mach numbers do not always have the same importance, and excluding one of them from consideration. For example, as can be seen from Fig. 3.7, the dimensionless drag force Π_x acting on a sphere in the region

$$Re > (Re)_{max} \approx 300\,000$$

becomes dependent on the Reynolds number only if the Mach number is less than Ma $\approx$ 0·8.* In general, the drag force F_x acting on an airfoil

* The curves in Fig. 3.7 were determined by A. Naumann[7,8].

need be considered as dependent on both Reynolds number and Mach number only under certain conditions; the lift force F_y being practically always independent of the Reynolds number (Fig. 3.8).* Accordingly,

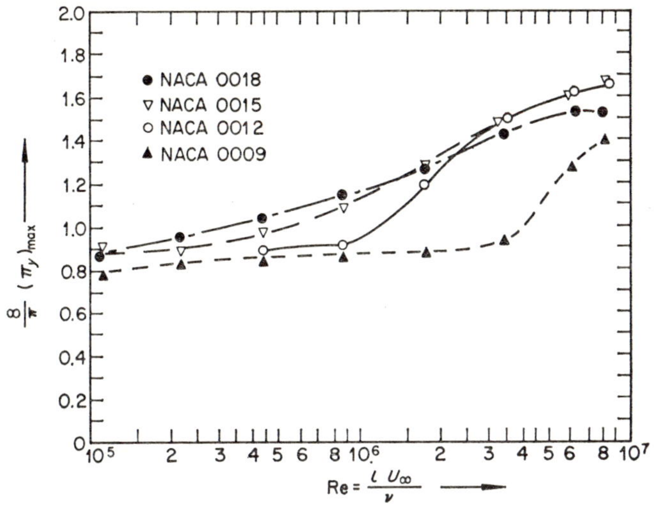

FIG. 3.8

(After A. M. Kuethe and J. D. Schetzer, *Foundations of Aerodynamics*, John Wiley, 1964)

when testing an airfoil for lift force, the eqn. (3.68) can practically always be treated as

$$\Pi_A \approx \varphi_A(\text{Ma}, \Gamma_1, \Gamma_2, \ldots) \tag{3.74}$$

When tests are carried out for drag force, the validity of (3.74) can be assumed only if the values of the Mach number are sufficiently large (supersonic aircraft). Since the property Π_A is given by (3.74) solely as a function of the Mach number and the geometry, the absolute size of the model, that is, the model scale λ_l, can be selected to some extent arbitrarily and thus the realisation of a geometrically similar small scale model (operating with the prototype speed of the airflow) becomes possible. From the eqn. (3.70) it follows that $\lambda_{\text{Re}} = \lambda_l$, i.e. the smaller the model, the larger is the discrepancy between the model and the prototype Reynolds numbers Re″ and Re′. One should remember that (3.74) is an approximate equality, and therefore, even if the influence of Re is small it exists. Hence, the bigger the model the nearer Re″ is to Re′, and so the nearer are the

* In Fig. 3.8 the subscript max in $(\Pi_y)_{\max}$ implies that the plotted values of Π_y correspond to those angles of attack which yield maximum value for the lift force.

model and prototype values of the property Π_A (which are actually given by the model and prototype versions of (3.68)) to each other.

Conversely, there are cases when the influence of the Mach number, though it exists, is negligible in comparison to that of the Reynolds number (subsonic aircraft), in which case (3.68) can be approximated by

$$\Pi_A \approx \varphi_A(\text{Re}, \Gamma_1, \Gamma_2, \ldots) \tag{3.75}$$

that is, by the relation studied so far in the preceding sections.

Certainly, there are cases when one cannot neglect the influence of either Re or Ma, and this is precisely the reason why such enormous wind tunnels are built (Plate 1) to test the aircraft models.

3.6 Cavitation

If, in a region occupied by the flow of a *liquid*, the absolute pressure drops to the value of the vapour pressure p_v, then a local vaporization of the

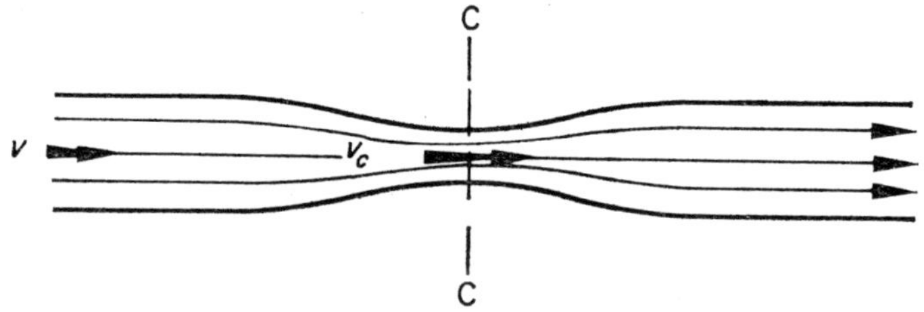

FIG. 3.9

liquid takes place in that region. Consider, for example, the flow in a closed conduit shown in Fig. 3.9. Let us assume that the flow rate Q, and thus the average velocity v_c, corresponding to the contracted section C—C, progressively increases. The increment of v_c will cause, according to the Bernouilli equation:

$$\frac{\alpha v_c^2}{2g} + \frac{p_c}{\gamma} + z_c = \text{const}$$

the decrement of the absolute pressure p_c of the section C—C. When the decreasing value of p_c becomes as small as p_v, that is, when the equality

$$p_c = p_v$$

is established, then the vapour bubbles (or cavities) appear in the section C—C. These vapour bubbles are carried by the flow downstream into the regions of smaller velocity, and thus of larger pressure (than $p_c = p_v$)

where they become condensed. The condensation of the bubbles takes place in an extremely short time. In other words, the liquid surrounding the bubbles rapidly moves towards its centre and suddenly stops there. This implies a practically instantaneous conversion of the kinetic energy of the liquid (moving towards the centre of the bubble) into pressure energy. The continuous occurrence of these sudden conversions of kinetic energy into pressure energy, for the multitude of bubbles, manifests in the form of an irregular sequence of loud bangs which is referred to as *cavitation*. In the design of hydraulic machinery, cavitation is of particular importance as it affects the metal in a mechanical as well as in a chemical sense (erosion and corrosion, respectively).

If, in the model, the prototype fluid is used, then (assuming the identity of the model and the prototype values of the temperature and barometric pressure) the study of cavitation in the model can be carried out only if the absolute pressure of the model flow is identical to that of the prototype flow, i.e. if

$$\lambda_p = 1 \tag{3.76}$$

On the other hand, since

$$\Pi_p = \rho^{-1} l^0 U^{-2} p = \frac{p}{\rho U^2} \tag{3.77}$$

we have

$$\lambda_p = \lambda_\rho \lambda_U^2 \tag{3.78}$$

From (3.76) and (3.78) it follows, however, that the realisation of a small scale model is impossible. Indeed, considering that $\lambda_\rho = 1$, and taking into account that in a Reynolds model the velocity scale is given by $\lambda_U = 1/\lambda_l$, from (3.76) and (3.78) we arrive immediately at $\lambda_l = 1$. However, cavitation usually takes place at large values of the velocity (and consequently of the Reynolds number). It is therefore very likely that the flow under investigation is independent of the Reynolds number. If this is so, then the condition $\lambda_U = 1/\lambda_l$ (which follows from the equality of the model and prototype values of this Reynolds number) can be relaxed, while (3.76) and (3.78) will yield merely $\lambda_U = 1$. In other words, if the flow is independent of the Reynolds number, then the study of cavitation on a small scale model becomes possible, provided the model has the same values of the velocity and pressure as the prototype. The situation becomes considerably more difficult if the equality of the model and prototype values of the Reynolds numbers cannot be relaxed, and it is still required that the model operates with the prototype fluid and is a small scale model.

A model subjected to such severe restrictions can be designed only according to the method explained in (ii) of the Section 2.4. Consider the prototype and model versions of the Bernouilli equation written for the section C—C

$$H_c' = \frac{v_c'^2}{2g} + \frac{p_c'}{\gamma} \tag{3.79}$$

and

$$H_c'' = \frac{v_c''^2}{2g} + \frac{p_c''}{\gamma} \tag{3.80}$$

It is assumed that the datum ($z = 0$) passes through the centroid of the section. Since

$$v_c'' = \lambda_U v_c' = \frac{1}{\lambda_l} v_c' \tag{3.81}$$

$$H_c'' = \lambda_H H_c'$$

while

$$p_c'' = p_c' \tag{3.82}$$

the eqn. (3.80) can be expressed as follows

$$H_c' = \frac{1}{\lambda_H}\left(\frac{1}{\lambda_l^2}\frac{v_c'^2}{2g} + \frac{p_c'}{\gamma}\right) \tag{3.83}$$

Equating (3.83) and (3.79) we obtain

$$\lambda_H = \frac{\dfrac{1}{\lambda_l^2}\dfrac{v_c'^2}{2g} + \dfrac{p_c'}{\gamma}}{\dfrac{v_c'^2}{2g} + \dfrac{p_c'}{\gamma}} \tag{3.84}$$

i.e.

$$\lambda_H = \frac{\dfrac{1}{\lambda_l^2} N' + 1}{N' + 1} \tag{3.85}$$

where N' is the prototype value of the dimensionless combination

$$N = \frac{1}{2}\frac{\rho v_c^2}{p_c} \tag{3.86}$$

Clearly, the relation between the scales λ_H and λ_l can also be given in terms of the model value N'' as follows:

$$\lambda_H = \frac{N'' + 1}{\lambda_l^2 N'' + 1} \tag{3.87}$$

Observe that for the small scale model, i.e. for $\lambda_l < 1$, we have $\lambda_H > 1$. The derived equations correspond to a model and prototype relation shown

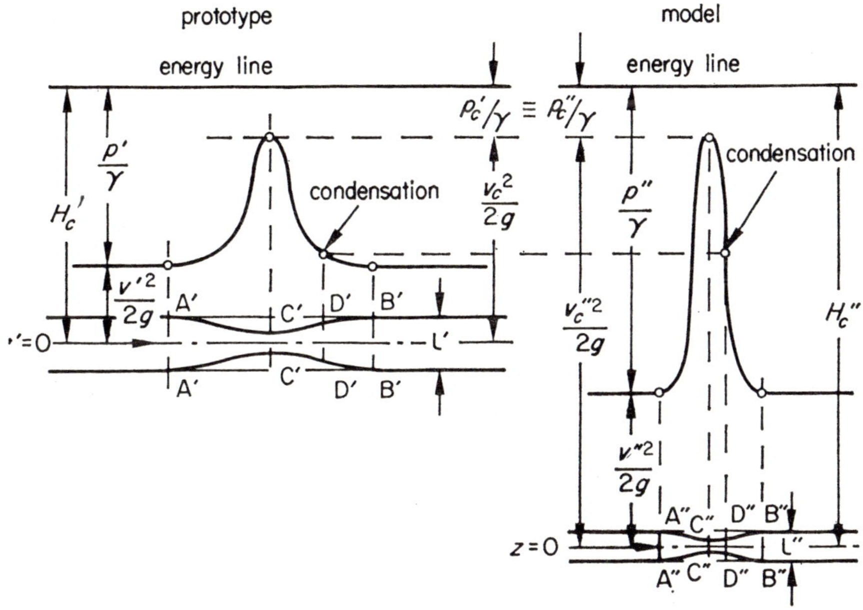

FIG. 3.10

schematically in Fig. 3.10. It is assumed that the energy loss due to friction (along a relatively short pipe) and that due to the gradual transition is negligible. Hence, the energy line in Fig. 3.10 is drawn simply as a horizontal straight line. The flow velocity of the small scale model is larger than that of the prototype as required by the Reynolds criterion. On the other hand, the absolute pressure at C′—C′ and C″—C″ is the same. Hence, the model head H_c'' is larger than the prototype head H_c' as required by (3.85) and (3.87). Any quantity determined in the model in relation to section C—C for a model value N'' will correspond to that prototype flow which is produced by the head

$$H_c' = H_c''/\lambda_H$$

where the scale λ_H (given by (3.87)) is dependent on the model value N'' under investigation. Hence, the prototype value H_c' is a function of both H_c'' and N''. A similar situation will arise if the prototype value of a quantity related to the section C—C is to be converted to the model value (in accordance with (3.85)). In this case, the model value of the converted quantity will correspond to such a model head H_c'', which is determined by both H_c' and N'.

The disadvantage of this method is the fact that the required equality of the absolute pressure is achieved only for the sections most important with respect to the cavitation, C′—C′ and C″—C″. As can be seen from the schematic diagram in Fig. 3.10, the absolute pressure in the model sections, other than C″—C″, is higher than that of the corresponding prototype sections. Thus, even if the model does reflect the origin of cavitation (in section C″—C″) in a dynamically similar manner, the subsequent destiny of the cavities dragged by the flow, will not strictly speaking, be similar; they will disappear, for example, in the model earlier than in the prototype (because the pressure required for the condensation will be present in the model in, say, a section D″—D″ that is relatively nearer to C″—C″ than the prototype condensation section D′—D′ is to C′—C′).

Sometimes the similarity in cavitation is considered by introducing the *cavitation parameter*[10]

$$K = \frac{p - p_v}{\frac{1}{2}\rho U^2} \tag{3.88}$$

where p and U are the typical pressure and typical velocity of the system. For example, p can be the pressure at the section A—A or at B—B etc. (Fig. 3.10), while U can be the average velocity v of the flow in the uniform part of the pipe. It is assumed that the similarity in cavitation is given by the identity of model and prototype values of the cavitation parameter, i.e. by the condition

$$\lambda_K = 1$$

which will obviously be satisfied by all kinematically similar systems, having identical ratio of all dimensionless velocities such as $v_c/U = v_c/v$. Let us suppose that the model and prototype fluids are the same, including the temperature and barometric pressure. This implies that the model and prototype values of p_v and ρ are identical. In this case the identity of K' and K'' implies

$$\frac{p' - p_v}{U'^2} = \frac{p'' - p_v}{U''^2}$$

i.e.

$$\lambda_U{}^2 = \frac{p'' - p_v}{p' - p_v}$$

Observe that the left hand side of this equation is constant for it is the power of a scale, whereas the right hand side varies with the stage. Hence, the method that rests on the equality of the cavitation parameter, strictly speaking, can provide similarity for one stage and one location only.

REFERENCES

1. V. A. Vanoni, N. H. Brooks; 'Laboratory Studies of the Roughness and Suspended Load of Alluvial Streams', *California Inst. Tech. Sedimentation Lab.*, Rept. No. E-68, 1957.
2. H. Schlichting, *Boundary Layer Theory*, McGraw-Hill, Book Co. Inc., Verlag G. Braun (Sixth Edition) (1968).
3. N. Webber, *Fluid Mechanics for Civil Engineers*, E. & F. N. Spon (1968).
4. G. K. Batchelor, *An Introduction to Fluid Dynamics*, Cambridge University Press (1967).
5. L. Prandtl, 'Der Luftwiderstand von Kugeln', *Nachr. Ges. Wiss. Göttingen*, Math. Phys. Klasse 1914, 177–190. See also: *Collected Works* II, 597–608.
6. A. Roshko, *J. Fluid Mechanics*, **10,** 345 (1961).
7. A. Naumann, 'Luftwiderstand von Kugeln bei hohen Unterschallgeschwindigkeiten', *Allgem. Wärmetechnik*, **4,** 217–221 (1953).
8. A. Naumann and H. Pfeifer, 'Über die Grenzschichtströmung am Zylinder bei hohen Geschwindigkeiten', *Adv. in Aero Sci.* (ed. Th. Von Kármán), **3,** 185–206 (1962).
9. A. M. Kuethe, J. D. Schetzer, *Foundations of Aerodynamics*, John Wiley, 2nd ed. (1964).
10. J. E. Plapp, *Engineering Fluid Mechanics*, Prentice-Hall (1968).

4 Flow Through Porous Media, Filtration Models

4.1 General

So far as the dimensional nature of characteristic parameters and the meaning of dimensionless variables is concerned, filtration flows are much the same as those forming the subject of the previous chapter. However, the physical picture of filtration flows and the fields of their application are so different from the 'free' flows that they are usually studied separately. In accordance with this convention it is intended to analyse filtration flows by introducing this separate chapter. When referring to filtration we refer to the motion of so-called *gravitational* liquid, continuously occupying the pores of the granular (or porous) medium. Unlike the 'hygroscopic water' or 'pellicular water', gravitational water (or, in general, fluid) is not subjected to the action of tension forces directed towards the surface of solid particles. Gravitational fluid, which forms the subject of filtration, moves, under natural conditions, because of gravity forces alone. (See the classification and the description of various components of water in soil.[1,2,3,4])

In the following pages we shall refer to the totality of a large number of solid grains as granular material, and to the space occupied by granular material as granular medium. The geometry of granular material and a granular medium is determined by the geometric properties listed at the top of the next page. If the geometric properties of granular medium do not vary (in the statistical sense) from one location in the medium to another, then the granular medium is said to be *homogeneous*; if they are independent of the direction, the medium is *isotropic*. The granular material, or medium, which is formed by loose, uncemented, grains is referred to as *cohesionless*. A granular medium which consists of a well-mixed, cohesionless, granular

(i) the shape of the grains (ii) the shape of the dimensionless grain size distribution curve	geometric properties of granular material	geometric properties of granular medium
(iii) the porosity coefficient, or the void ratio n (implying the ratio of the void part of a volume of granular medium to the total volume)		

material can always be regarded as homogeneous and isotropic, and its grains are packed in their own (natural) way so as to yield a certain (natural) value of the porosity coefficient n. In such cases the porosity coefficient n can be considered as a quantity determined by the geometric properties (i) and (ii) of granular material; this is why sometimes one can associate a certain sand, corresponding to a certain geometry (i) and (ii), with a certain value of porosity. (See Table 3 in Ref. 1.) However, n cannot be regarded as determined by (i) and (ii) alone if the medium is not pure, or if it is not cohesionless, for the cementing material can occupy a considerable percentage of the space between the grains, and thus it can reduce the value of n considerably.

If the geometry of a granular medium is specified, then the absolute size of its grains can be defined by the numerical value of a single typical grain diameter (selected for the comparison of absolute sizes). Since the absolute size can be 'measured' by any 'length unit', any grain diameter, corresponding to any percentage of the grain size distribution curve, can be selected as typical. Accordingly, any one of, say, D_{50}, D_{90}, D_{max} etc. can be used as the typical (or representative) diameter. In order to characterize this freedom in selection, in the following, the typical diameter will be denoted simply by D (without any subscript). Attention is drawn to the fact that the typical grain diameter is not a parameter determining the geometry; it is a parameter determining the absolute size of the grains forming the granular medium having the (internal) geometry that is determined by (i), (ii) and (iii) (independently of the absolute size). Depending on the absolute size of the grains, the granular medium is referred to by different names. Thus, *clay, silt, sand, gravel, pebble, boulder*, are different names given to the granular medium depending on the absolute size of its grains. In the theory of dimensions and similarity, the absolute value of a dimensional quantity (such as the 'length' D) is of no importance. The mechanics of the fluid motion in porous medium, due to the forces of gravity, is not determined by the absolute value of the grain size D. It is determined (as will be seen later) by the value of the Reynolds number $vD\rho/\mu$ which can possess the *same* value for *various* values of D. In order to emphasize this important fact, we will refrain from the use of names associated with the absolute value of a dimensional parameter, and use only the terms granular medium or porous medium. The ranges of D corresponding to clay, silt, sand, etc. can be found elsewhere[1,2].

4.2 Filtration Law

From the considerations in the first chapter (Fig. 1.5e), it is clear that any dimensionless property Π_A of the flow through a porous medium of a specified geometry must be a certain function of the Reynolds number and of the ratio of the grain diameter to the conduit size. Being now concerned with the filtration law itself, we assume that the size of the conduit is so large (in comparison to the size of the grains) that it has no influence on the progress of filtration taking place in the midst of the porous medium. Thus, we are left with the relation

$$\Pi_A = \bar{\varphi}_A \left(\frac{vD\rho}{\mu} \right) \tag{4.1}$$

where v is the filtration velocity and D the selected typical grain diameter of the porous medium. The form $\bar{\varphi}_A$ of the function above is dependent on the nature of the property A under investigation and on the geometric properties (i), (ii) and (iii) of the granular medium. Introducing the porosity n as an additional dimensionless variable we can express (4.1) in the following manner

$$\Pi_A = \varphi_A \left(\frac{vD\rho}{\mu}, n \right) \tag{4.2}$$

where the form of φ_A is dependent now only on the nature of the property A and on the geometric properties (i) and (ii) of the granular material.

Consider the Bernouilli trinomial

$$H = \frac{v^2}{2g} + \frac{p}{\gamma} + z \tag{4.3}$$

corresponding to any point in the porous medium. In the filtration flow the velocity head $v^2/2g$ is always negligible in comparison to the pressure head (p/γ), and thus the energy gradient can always be expressed as follows:

$$J = -\frac{\mathrm{d}H}{\mathrm{d}x} = -\frac{\mathrm{d}\left(\dfrac{p}{\gamma} + z\right)}{\mathrm{d}x} \tag{4.4}$$

where x is the direction of the flow (which can be straight as well as curvilinear). The fact that $v^2/2g$ is negligible implies that the energy gradient $-\mathrm{d}H/\mathrm{d}x$ (of a gradually varying filtration flow) coincides with the slope of the free surface (if it exists).

Let the property A under investigation be the gradient of the total pressure $p + \gamma z$. We have

$$A = \gamma J \tag{4.5}$$

Using (as in the previous chapter) ρ, v and D as basic quantities, we obtain

$$\Pi_J = \rho^{-1} v^{-2} D^1 \gamma J = \frac{\gamma J D}{\rho v^2} \tag{4.6}$$

and thus

$$\Pi_J = \frac{\gamma J D}{\rho v^2} = \varphi_J(\mathrm{Re}, n) \tag{4.7}$$

where

$$\mathrm{Re} = \frac{v D \rho}{\mu} \tag{4.8}$$

Consider the following special cases.

(i) *Small values of* Re (laminar or viscous filtration)
In this case the contribution of the inertia forces on the formation of the energy gradient J is negligible in comparison to that of the viscous friction forces. Thus, the parameter ρ must vanish from the relation (4.7) but μ must remain. It is obvious that such a requirement can be fulfilled only if the function in (4.7) has the following form:

$$\varphi_J(\mathrm{Re}; n) = \frac{\varphi_1(n)}{\mathrm{Re}} \tag{4.9}$$

Thus, for small values of Re the law of filtration can be expressed as

$$\Pi_J = \frac{\gamma J D}{\rho v^2} = \frac{\varphi_1(n)}{\dfrac{v D \rho}{\mu}} \tag{4.10}$$

or, for example, as

$$v = \left[\frac{1}{\varphi_1(n)} \frac{\gamma D^2}{\mu}\right] J \tag{4.11}$$

which is nothing else but the well-known Darcy law

$$v = kJ \tag{4.12}$$

where k is the permeability coefficient

$$k = \left[\frac{1}{\varphi_1(n)} \frac{\gamma D^2}{\mu}\right] \tag{4.13}$$

The above considerations demonstrate how effective the dimensional methods can be. Indeed, the Darcy law, the expression of permeability coefficient, and the fact that the Darcy law can be valid only for small values of Re, have been derived here quite briefly by dimensional considerations. These notions became known in hydraulics (before the days of the theory of dimensions) as the cumulative result of the experimental investigations carried out for many years by a number of authors.

(ii) *Large values of* Re (turbulent filtration)
In this case the influence of the viscous forces is negligible in comparison to that of the inertia forces, and thus the parameter μ must vanish from the relation (4.7). Since μ appears only in the expression of the Reynolds number Re, this number must not be present in the law of turbulent filtration. Thus,

$$\varphi_J(\mathrm{Re}; n) = \varphi_2(n) \tag{4.14}$$

and consequently

$$\Pi_J = \frac{\gamma J D}{\rho v^2} = \varphi_2(n) \tag{4.15}$$

The graph in Fig. 4.1 shows the experimental values of the dimensionless combination Π_J plotted versus the Reynolds number Re for various granular materials possessing various values of the porosity n.* From this plot in log log systems of co-ordinates, it may be seen that the experimental curves become indistinguishable from the 45° straight lines (implying the eqn. (4.10)) when the order of Re becomes less than unity. Similarly, the curves appear to become parallel to the Re axis as implied by (4.15) when the order of Re exceeds $\approx 10^4$. Hence the orders ≈ 1 and $\approx 10^4$ can be adopted as a rule of thumb in order to distinguish 'small' and 'large' values of Re. Thus

$$\left.\begin{array}{l}\mathrm{Re} < \approx 1 \text{ (small values of Re—\textit{laminar} filtration)}\\ \mathrm{Re} > \approx 10^4 \text{ (large values of Re—\textit{turbulent} filtration)}\end{array}\right\} \tag{4.16}$$

* In Fig. 4.1, the bar over $\overline{\Pi}_J$ and $\overline{\mathrm{Re}}$ signifies that these quantities are formed by the 'effective grain diameter' $\overline{D}$ (which will be defined presently).

1 Wind Tunnel of The National Research Council of Canada, Ottawa

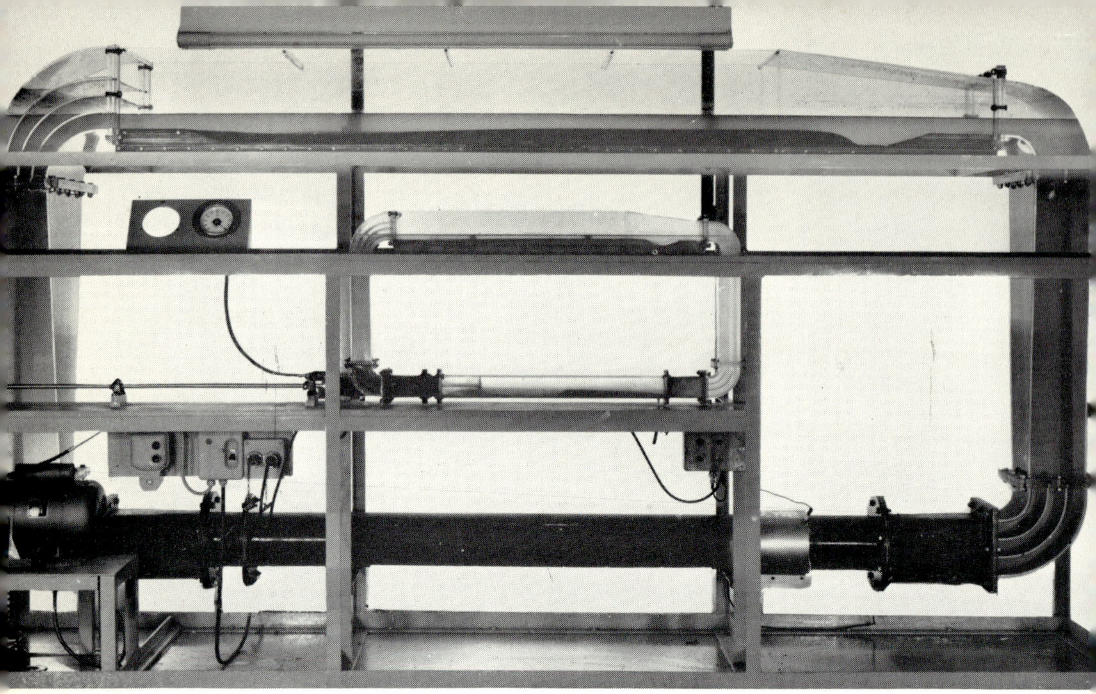

2 (b) 'Similarity Flumes', Hydraulic Research Station, Wallingford

2 (a) Roughness of the model of the Mosel River at Enkirch

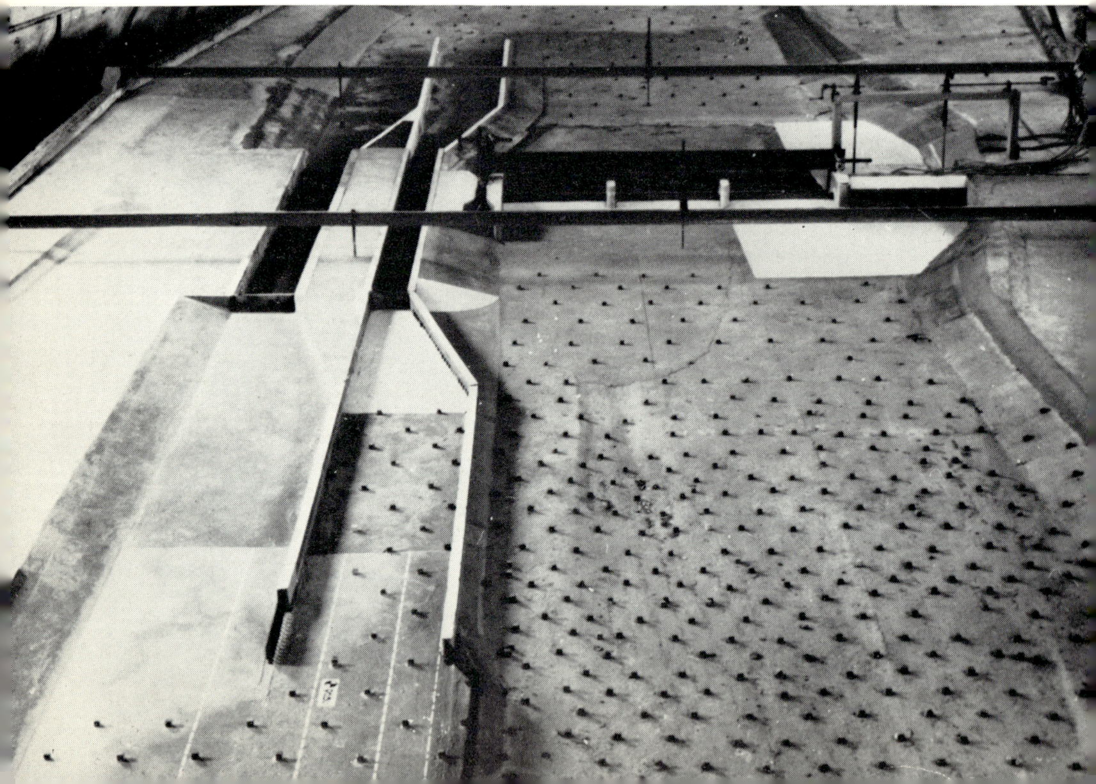

Sample No.	Sand	Porosity
	Consolidated	%
1	Bradford	12·5
2	Bradford	12·3
3	3rd Venango	16·9
4	Ceramic A	37·0
5	Robinson	20·3
6	Ceramic B	37·8
7	Woodbine	19·7
8	Wilcox	15·9
9	3rd Venango	11·9
10	Robinson	19·5
11	Robinson	18·4
12	3rd Venango	22·3
13	Wilcox	16·3
14	Warren	19·2
15	3rd Venango	21·4
16	Robinson	20·6
17	Ceramic C	33·2
18	3rd Venango	21·9
19	Woodbine	23·8
20	,,	26·9
21	,,	27·7
22	,,	22·1
23	,,	28·8
	Unconsolidated	
24	Flint	38·5
25	Ottawa	30·9
26	20-30 Ottawa	34·5
27	Lead shot	34·5

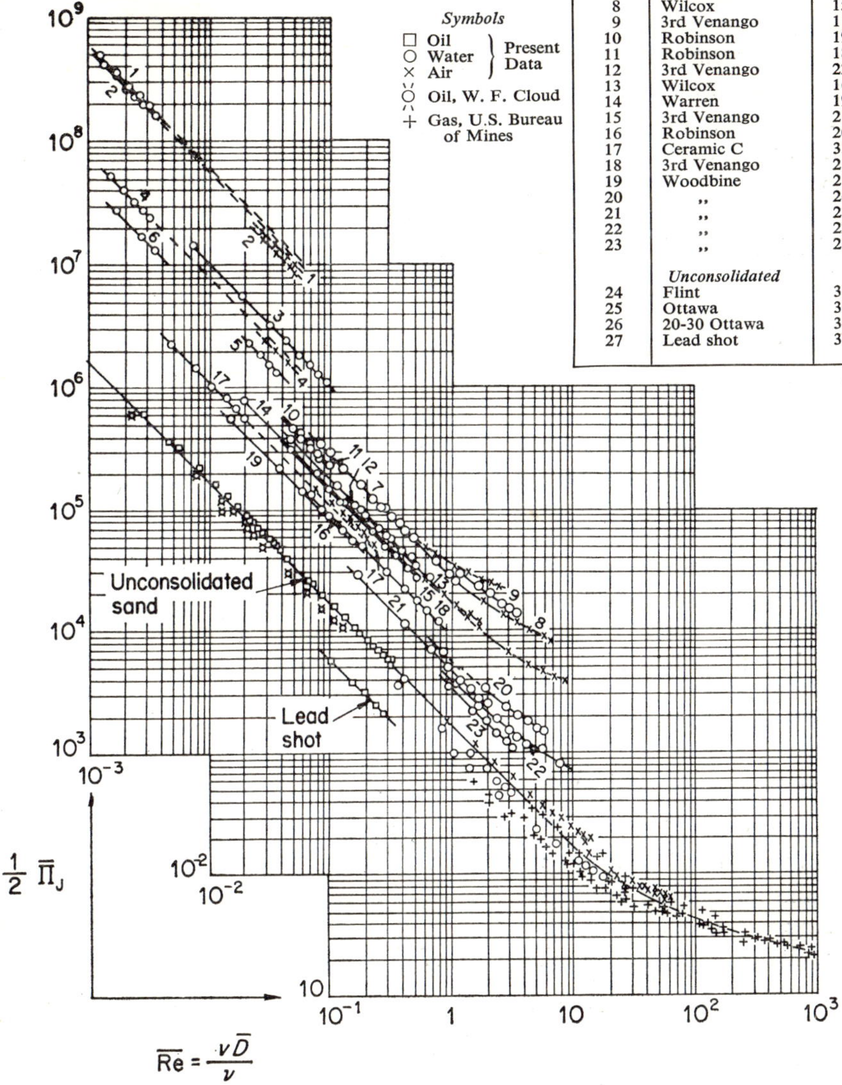

FIG. 4.1

Friction-factor chart for the flow of fluids through sands. (After Fancher, Lewis and Barnes, Bull. Pa. State Expt. Sta.)

It follows that the *transitional* interval

$$\approx 1 < \text{Re} < \approx 10^4 * \tag{4.17}$$

is considerable; in practice, many cases fall into this interval. It seems that the popularity of the Darcy law in the treatment of filtration problems is mainly because of its mathematical convenience (*linear* Laplace differential equation, potential flow theory), rather than because it covers the majority of practical cases.

At this stage we introduce the concept of *effective diameter*. The form of all functional relations used since the beginning of this chapter is supposed to be dependent on the geometric properties (i) and (ii) of granular material. In other words, the form of the mentioned functions is itself a function of the factors (i) and (ii). Since the early days of research on filtration, attempts have been made to eliminate the consideration of geometric properties (i) and (ii) by introducing an appropriately selected 'effective grain diameter', which will be denoted henceforward by $\bar{D}$.

From the physical point of view the effective diameter $\bar{D}$ is the answer to the following question. 'Given a nonuniform granular medium (consisting of non-equal size grains), and having a certain porosity n, what is the size $\bar{D}$ of a fictitious uniform medium, consisting of equal size spherical grains, which possesses the same value of the porosity n, and which, under the action of the same gradient J, yields for the same fluid, the same value of filtration velocity v (as the given non-uniform medium)?' Let φ_J and $\bar{\varphi}_J$ be the functions corresponding to the actual medium and the fictitious medium respectively. Using this notation, the effective diameter can be defined (in a mathematical manner) as a quantity satisfying the following equation

$$\varphi_J\left(\frac{vD\rho}{\mu}, n\right) = \bar{\varphi}_J\left(\frac{v\bar{D}\rho}{\mu}, n\right)$$

The concept of effective diameter which implies the substitution of three factors (i), (ii) and D by only one quantity $\bar{D}$ is undoubtedly very attractive, and its application in practice may in certain circumstances be rewarding. On the other hand, such a concept can hardly be regarded as sound from a rigorous scientific point of view. Indeed, as is clear from the

* It is difficult to determine when a curve becomes indistinguishable from a straight line. There is no consistency in the literature with respect to the values ≈ 1 and $\approx 10^4$ (representing the limits of the transitional region). For example, the lower limit of the transitional regime (the upper limit of the Darcy law) according to Ref. 1 is 'a number between 3 and 10', according to Ref 5 it is 6, according to Ref. 2 it is between 1 and 10. We adopt the most severe (with respect to the application of Darcy law), and simplest to remember value ≈ 1. The same value is adopted by G. K. Batchelor[6]. Similar comments are valid with respect to $\approx 10^4$.

equality above, the concept of effective diameter essentially says that the curve

$$\bar{\Pi}_J = \bar{\varphi}_J(\text{Re}, \text{const})$$

which corresponds to the uniform medium consisting of spherical grains (and which represents a certain value $n = \text{const}$) can always be made identical to any curve

$$\Pi_J = \varphi_J(\text{Re}, \text{const})$$

merely by multiplying the abscissa Re and the ordinate $\bar{\Pi}_J$ by appropriately selected constants. From the available experimental curves representing the variation of any quantity with any kind of the Reynolds number, it is clear that the curves corresponding to different geometry of the same phenomenon cannot be made congruent simply by shifting them in a log log system of coordinates. In general, the identity of the curves can be achieved for some limited regions only, where the curves (plotted on log log coordinates) become the straight lines (corresponding to small and large values of Re). Accordingly, the concept of the effective diameter, strictly speaking, can be valid only when $\text{Re} < \approx 1$ (Darcy law region), i.e. when the curve Π_J becomes the 45° straight line, or when it becomes undistinguishable from the straight line parallel to abscissa ($\text{Re} > \approx 10^4$). However, since all Π_J curves begin as 45° straight lines, and end as the straight lines parallel to the abscissa, they are reasonably similar to each other, and therefore, the error cannot be very large (at least not for practical purposes) if it is assumed that they can be made congruent by a suitable choice of $\bar{D}$. Considering this, and taking into account the advantages provided by the simplicity of this conventional effective diameter method, the granular material will be defined, henceforth, by its effective diameter $\bar{D}$, while the granular medium will be given by the pair $\bar{D}$ and n. There are various methods of determining the effective diameter corresponding to the given granular material which will not be considered here, and the reader interested in them is referred to the special literature on filtration.* In the following part of the present chapter, the bar will be used only in order to signify the quantities corresponding to the effective grain diameter $\bar{D}$.

The curves shown in Fig. 4.1 correspond to a variety of granular materials, and to all possible geometric properties (i) and (ii). The results of measurements plotted in Fig. 4.1 were not subjected to any classification with respect to the properties (i) and (ii), they were plotted simply by taking into account the value of the effective diameter $\bar{D}$ (when computing

* See the methods for determining $\bar{D}$ (suggested by F. King, A. Hazen, Krüger-Zunker, J. Kozeny and E. A. Zamarin) in Chapter XXVI, Ref. 8.

Re and $\overline{\Pi}_J$), and the porosity n. In other words, it was assumed that the granular medium can be satisfactorily described by the numerical values of the parameters $\bar{D}$ and n alone.

The curve family in Fig. 4.1, that is, the function of two variables $\overline{\Pi}_J = \bar{\varphi}_J(\overline{\text{Re}}, n)$ can be reasonably well represented by the following form (which no longer depends on (i) and (ii))

$$\tfrac{1}{2}\overline{\Pi}_J = \frac{1}{n^6}\left(0{\cdot}01 + \frac{1}{\overline{\text{Re}}}\right) \tag{4.18}$$

This relationship, which implies that the energy gradient J is given by a second order function of v (i.e. as $J = av^2 + bv$) is essentially the same as the quadratic forms used by E. Lindquist[9], G. Schneebeli[10], P. Nemenyi[11], S. A. Khristianovich[12], and others. The relationship (4.18) differs from those of the authors mentioned because, in addition to Re (i.e., v) it indicates explicitly the influence of the porosity n on the law of filtration. It is not yet certain, however, whether $\overline{\Pi}_J$ varies exactly with the *sixth* power of n. For example, G. Cohen De Lara[13] suggests the *fifth* power ($J \sim 1/n^5$). A review of the expressions suggested for the regions of non-linear filtration (outside of the region of validity of the Darcy law) can be found[1,2,3]. Accordingly, the diagram in Fig. 4.1 can be superceded by a more systematic, and convenient diagram in Fig. 4.2.

4.3 Similarity Criteria for General Cases of Filtration Flow

From (4.7), i.e. from

$$\overline{\Pi}_J = \bar{\varphi}_J(\overline{\text{Re}}; n)$$

(where the function is expressed by the effective diameter) it follows that the dynamic similarity of filtration is given by

$$\lambda_{\overline{\text{Re}}} = \lambda_v \lambda_{\bar{D}} \lambda_\rho \lambda_\mu{}^{-1} = 1 \tag{4.19}$$

and

$$\lambda_n = 1 \tag{4.20}$$

From the expressions above, it is clear that in principle the model for filtration is indeed a Reynolds model, as pointed out at the beginning of the chapter. However, the design of a filtration model, is usually, more complicated than that of an ordinary Reynolds model. Indeed, in the majority of filtration problems the *free surface* is involved; either the filtration flow itself has a free surface, or the 'reservoirs' which generate

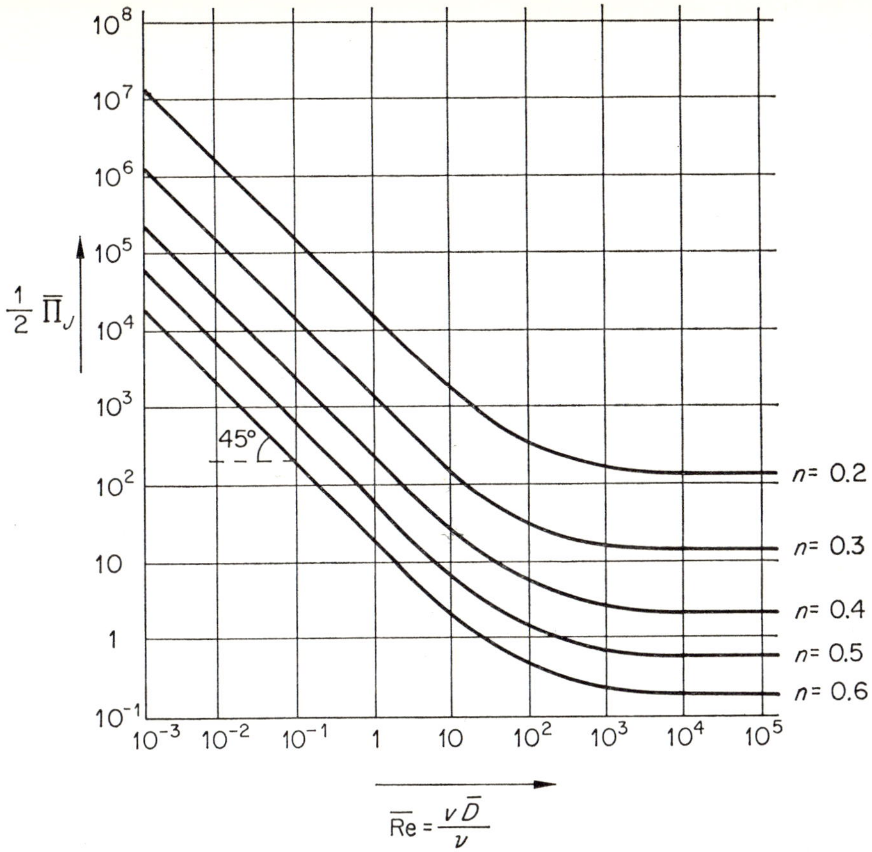

FIG. 4.2

the filtration flow have free surfaces of different levels. In either of these cases, the free surface must be reproduced in the model in an undistorted (geometrically similar) manner. Thus, in the case of a filtration model it is not sufficient to have

$$\lambda_{\overline{\Pi}_J} = 1 \quad \text{i.e.} \quad \lambda_\gamma \lambda_J \lambda_{\overline{D}} \lambda_\rho^{-1} \lambda_v^{-2} = 1 \tag{4.21}$$

alone; one has to satisfy in addition

$$\lambda_J = 1 \tag{4.22}$$

If the model operates with the fluid of the prototype (water), i.e. if

$$\lambda_\rho = \lambda_\mu = \lambda_\gamma = 1 \tag{4.23}$$

then (4.19) and (4.21) become

$$\lambda_v = \frac{1}{\lambda_{\bar{D}}} \tag{4.24}$$

and

$$\lambda_v^2 = \lambda_J \lambda_{\bar{D}} \tag{4.25}$$

respectively. From these expressions, it follows clearly that $\lambda_J = 1$ implies $\lambda_{\bar{D}} = 1$ and $\lambda_v = 1$, and vice versa. Attention is drawn to the fact that $\lambda_{\bar{D}}$ is the scale of the effective diameter, not the scale of the model! Accordingly, $\lambda_{\bar{D}} = \lambda_v = 1$ must not necessarily be taken as implying that a small scale model is impossible. As will be seen later, in some cases it is perfectly possible to have $\lambda_{\bar{D}} = \lambda_v = \lambda_J = 1$ in a small scale model. However, in other cases, it is not possible, and therefore, it is desirable to find a more general method which would include $\lambda_{\bar{D}} = \lambda_v = \lambda_J = 1$ as a special case.

We adopt the approach explained in paragraph (ii) of Section 2.4 and consider the expression (4.18) for $\overline{\Pi}_J$. The prototype and model versions of (4.18) are

$$\tfrac{1}{2}\overline{\Pi}_J' = \frac{1}{n'^6}\left(0{\cdot}01 + \frac{1}{\overline{\mathrm{Re}}'}\right)$$

and (4.26)

$$\tfrac{1}{2}\overline{\Pi}_J'' = \frac{1}{n''^6}\left(0{\cdot}01 + \frac{1}{\overline{\mathrm{Re}}''}\right)$$

respectively. Dividing these equations we obtain

$$\frac{\overline{\Pi}_J''}{\overline{\Pi}_J'} = \left(\frac{n'}{n''}\right)^6 \left[\frac{0{\cdot}01 + \dfrac{1}{\overline{\mathrm{Re}}''}}{0{\cdot}01 + \dfrac{1}{\overline{\mathrm{Re}}'}}\right] \tag{4.27}$$

and since

$$\lambda_{\overline{\Pi}_J} = 1$$

must be valid

$$\left(\frac{n''}{n'}\right)^6 = \frac{0{\cdot}01 + \dfrac{1}{\overline{\mathrm{Re}}''}}{0{\cdot}01 + \dfrac{1}{\overline{\mathrm{Re}}'}}$$

i.e.

$$\lambda_n^{\,6} = \frac{0{\cdot}01\overline{\mathrm{Re}}' + \lambda_{\overline{Re}}^{-1}}{0{\cdot}01\overline{\mathrm{Re}}' + 1} \tag{4.28}$$

In the case of the model operating with the prototype fluid, we have (as is clear from (4.19))

$$\lambda_{\overline{Re}} = \lambda_v \lambda_{\bar{D}} \tag{4.29}$$

and therefore (4.28) can be expressed in the following manner

$$\lambda_n^{\,6} = \frac{0{\cdot}01\overline{\mathrm{Re}}' + [\lambda_v \lambda_{\bar{D}}]^{-1}}{0{\cdot}01\overline{\mathrm{Re}}' + 1} \tag{4.30}$$

The equality of the model and prototype values of the energy gradient (free surface slope) J, implied by (4.22) has not yet been used. From (4.21) or (4.25), it is clear that the simultaneous validity of

$$\lambda_{\overline{\Pi}_J} = 1 \quad \text{and} \quad \lambda_J = 1$$

can be achieved only if

$$\lambda_v^{\,2} = \lambda_{\bar{D}} \tag{4.31}$$

Hence, the scales λ_n, λ_v and $\lambda_{\bar{D}}$ must be selected so as to satisfy both (4.30) and (4.31). In a model designed in this way, the friction factor $1/c^2$, which is proportional to $\overline{\Pi}_J$ as

$$\frac{1}{c^2} = \frac{v_*^{\,2}}{v^2} \sim \frac{gJ\bar{D}}{v^2} = \frac{\gamma J\bar{D}}{\rho v^2} = \overline{\Pi}_J \tag{4.32}$$

will be identical to that of the corresponding points of the prototype, while the field of the filtration velocities v'', corresponding to the given prototype value of the Reynolds number Re′, will be dynamically (or to be more precise, kinematically) similar to the prototype field. The validity of $\lambda_J = 1$ implies that the slope of the free surface of the filtrating flow (if it exists) does not undergo distortion.

Let us see the application of this method in some examples.

EXAMPLE 4.1

Consider the seepage flow from the channel A to the river B (Fig. 4.3). The granular medium marked by 1 is given by

$$n_1' = 0{\cdot}327 \qquad \bar{D}_1' = 10\ \text{mm} = 0{\cdot}01\ \text{m}$$

while the medium marked by 2 (below the line $\overline{bc}$) by

$$n_2' = 0{\cdot}250 \qquad \bar{D}_2' = 1\ \text{mm} = 0{\cdot}001\ \text{m}$$

It is desired to reveal the kinematics of filtration flow by means of a model. In order to design the model according to the eqns. (4.30) and (4.31), one must determine first the orders of the prototype Reynolds numbers $\overline{\mathrm{Re}}_1'$ and $\overline{\mathrm{Re}}_2'$. Since the medium 2 is considerably less permeable than the

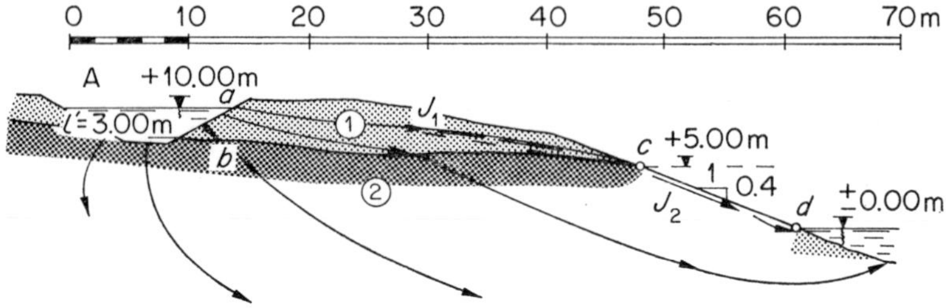

FIG. 4.3

medium 1 ($\bar{D}_2' = 1/10\bar{D}_1'$) the free surface of the flow percolating through the medium 1 will almost certainly extend up to the point c, and thus

$$J_1' \approx 0{\cdot}1$$

can be adopted. On the other hand, the flow in the medium 2 will approximately follow $\overline{\mathrm{cd}}$ (even if $\overline{\mathrm{cd}}$ is protected by a porous protection). Accordingly,

$$J_2' \approx 0{\cdot}4$$

can be taken with respect to the flow in ground 2. Substituting $n_1' = 0{\cdot}327$, $\bar{D}_1' = 0{\cdot}01$ m and $J_1' = 0{\cdot}1$ in (4.18), i.e. in

$$\frac{gJ\bar{D}}{2v^2} = \frac{1}{n^6}\left(0{\cdot}01 + \frac{\nu}{v\bar{D}}\right)$$

and considering that $g = 9{\cdot}81$ m/s² while $\nu = 10^{-6}$ m²/s (water in ≈ 20°C) we obtain

$$\frac{9{\cdot}81 \times 0{\cdot}1 \times 0{\cdot}01}{2v^2} = \frac{0{\cdot}01}{(0{\cdot}327)^6}\left(1 + 100\,\frac{10^{-6}}{v \times 0{\cdot}01}\right)$$

i.e.

$$2v^2 + 0{\cdot}01v - 0{\cdot}0012 = 0$$

which gives

$$v = v_1' = 0{\cdot}0221 \text{ m/s}$$

and consequently

$$\overline{\mathrm{Re}}_1' = \frac{0{\cdot}0221 \times 0{\cdot}01}{10^{-6}} = 221$$

Similarly, substituting $n_2' = 0{\cdot}250$, $\bar{D}_2' = 0{\cdot}001$ m and $J_2' = 0{\cdot}4$ in the same relation, we obtain

$$\frac{9{\cdot}81 \times 0{\cdot}4 \times 0{\cdot}001}{2v^2} = \frac{0{\cdot}01}{(0{\cdot}250)^6}\left(1 + 100\,\frac{10^{-6}}{v \times 0{\cdot}001}\right)$$

i.e.

$$2v^2 + 0{\cdot}1v - 0{\cdot}000094 = 0$$

which gives

$$v = v_2' \approx 0{\cdot}001 \text{ m/s}$$

and

$$\overline{\mathrm{Re}}_2' = \frac{0{\cdot}001 \times 0{\cdot}001}{10^{-6}} \approx 1$$

Using the determined values of Re_1' and Re_2' in (4.30), and taking into account the condition (4.31) we arrive at the following system of equations

$$\left.\begin{aligned} \lambda_{n_1}{}^6 &= \frac{2{\cdot}21 + \lambda_{\bar{D}}{}^{-\frac{3}{2}}}{2{\cdot}21 + 1} \\ \lambda_{n_2}{}^6 &= \frac{0{\cdot}01 + \lambda_{\bar{D}}{}^{-\frac{3}{2}}}{0{\cdot}01 + 1} \\ &\text{and} \\ \lambda_v{}^2 &= \lambda_{\bar{D}} \end{aligned}\right\} \qquad (4.33)$$

There are three equations and four scales $\lambda_{\bar{D}}$, λ_v, λ_{n_1} and λ_{n_2}, and thus only one can be selected freely. For the case under consideration, perhaps the most reasonable decision would be to choose the following simplest solution of the system (4.33)

$$\lambda_{n_1} = \lambda_{n_2} = \lambda_{\bar{D}} = \lambda_v = 1 \qquad (4.34)$$

and achieve a complete dynamic similarity of the filtration flow under investigation (i.e. with respect to all properties and stages). Such a solution would imply that the model media are formed by the prototype granular materials, while the filtration of the same fluid (water) is identical in the corresponding points of model and prototype (identical velocities, accelerations, friction forces, etc.). Let l be any external dimension of the cross-section shown in Fig. 4.3. Since the dimensionless combination

$$X = l/\bar{D}$$

does not affect the law of filtration, the condition $\lambda_{l/\bar{D}} = 1$, i.e. $\lambda_l = \lambda_{\bar{D}}$ can be relaxed. Thus, the model can possess any dimension l'' as long as

$l''/\bar{D}''$ is not so small as to affect the homogenuity of filtration. Selecting the distance (layer thickness) $\overline{ab}$ in the model as $l'' = 150$ mm, (i.e. as consisting of 15 grain diameters) and considering that the corresponding distance in the prototype (as seen from Fig. 4.3) is $l' = 3{\cdot}00$ m, we arrive at the geometric model scale

$$\lambda_l = \frac{l''}{l'} = \frac{15}{300} = \frac{1}{20}$$

Hence, the 70 m long prototype cross-section shown in Fig. 4.3 will be reduced in the model into 70/20 = 3·50 m.

Let us now assume that the model scale $\lambda_l = 1/20$ is satisfactory, however, one would prefer to have

$$\lambda_{\bar{D}} = 1/2 \text{ rather than } \lambda_{\bar{D}} = 1$$

Substituting $\lambda_{\bar{D}} = 1/2$ in (4.33), we obtain

$$\lambda_{n_1}{}^6 = \frac{2{\cdot}21 + 2^{\frac{3}{2}}}{2{\cdot}21 + 1} = \frac{5{\cdot}83}{4} = 1{\cdot}57, \quad \text{and thus } \lambda_{n_1} = 1{\cdot}078$$

$$\lambda_{n_2}{}^6 = \frac{0{\cdot}01 + 2^{\frac{3}{2}}}{0{\cdot}01 + 1} = \frac{2{\cdot}84}{1{\cdot}01} = 2{\cdot}81, \quad \text{and thus } \lambda_{n_2} = 1{\cdot}188$$

$$\lambda_v{}^2 = 1/2 \cdots\cdots\cdots\cdots , \text{ and thus } \lambda_v = 1/\sqrt{2} = 0{\cdot}707$$

In this case, the model granular material is determined by

$$\bar{D}_1'' = \tfrac{1}{2} \times 10 = 5 \text{ mm}; \quad n_1'' = 1{\cdot}078 \times 0{\cdot}327 = 0{\cdot}352$$

and

$$\bar{D}_2'' = \tfrac{1}{2} \times 1 = 0{\cdot}5 \text{ mm}; \quad n_2'' = 1{\cdot}188 \times 0{\cdot}250 = 0{\cdot}294$$

Observe how accurate the preparation of the model granular medium must be; when $\lambda_{\bar{D}}$ varies from 1 to 1/2, then the porosity scales vary from 1 to only 1·078 and 1·188, and this must certainly be observed, otherwise one cannot be sure that the filtration velocities of the layers 1 and 2 will be scaled down in the same ratio $1/\sqrt{2}$. Clearly, the model implied by the version (4.34) is preferable, not only because it is dynamically similar, and thus valid for all properties and stages, but also because it is easier to build (one uses simply the prototype material) and its design does not involve any estimations (like that of J_1' and J_2', i.e. of Re_1' and Re_2'). However, conditions might not always permit the realization of such a model, and therefore (as will be seen in the next example) one might have no alternative but to design the model according to eqns. (4.30) and (4.32).

EXAMPLE 4.2

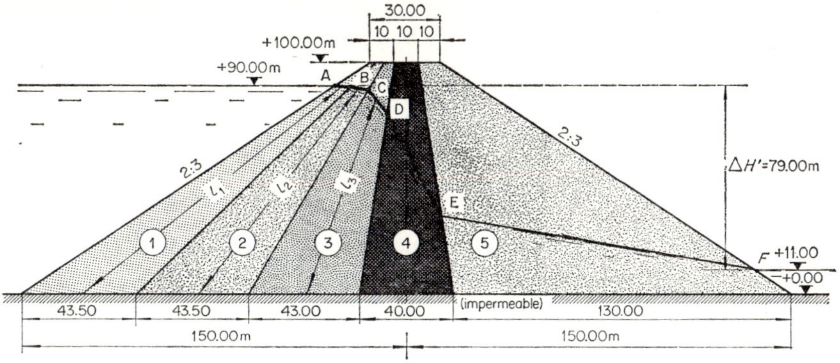

FIG. 4.4

Consider the heterogeneous earth dam shown in Fig. 4.4. The properties of its five zones are given below.*

Zone No. (i)	1	2	3	4	5
$\bar{D}_i'$ (mm)	100·0	50·0	10·0	7·50	50·0
n_i'	0·20	0·17	0·17	0·18	0·25

It is desired to determine the kinematic properties of the filtration flow corresponding to the highest water level difference $\Delta H' = 79$ m by model tests. In order to design the model, we begin again with the determination of the orders of prototype Reynolds numbers, $\overline{\text{Re}}_i'$ $(i = 1, 2, \ldots 5)$. These orders cannot be determined as simply as in the previous example. Indeed, they have to be determined by a trial and error method which consists of the following procedure:

1. One estimates the slope J_1' of the free surface in Zone 1, and knowing $\bar{D}_1'$ and n_1' from (4.18), one computes v_1' (the average value of filtration velocity in Zone 1).
2. Knowing J_1', one determines the position of the point B (see Fig. 4.4). From the point B one can draw a number of free surface slopes J_2' and for each of them determine the corresponding order of the filtration velocities as

$$v_2' = v_1' \frac{l_1'}{l_2'}$$

* No conclusions should be drawn from Fig. 4.4 and from the values in the Table with respect to the constructions of earth dams; the zones in Fig. 4.4 and the values $\bar{D}_i'$ and n_i are hypothetical.

the meaning of l_1' and l_2' being clear from Fig. 4.4. However, among these pairs $[J_2'; v_2']$ *only one* will satisfy (4.18).

3. Using this particular pair $[J_2'; v_2']$, one determines, in a completely analogous manner, the pair $[J_3'; v_3']$ (where $v_3' = v_2'l_2'/l_3'$), then $[J_4'; v_4']$, and so on.
4. The slope J_5' of the last determined pair $[J_5'; v_5']$ drawn from the point E must *pass through the fixed point* F, of intersection of the downstream slope of the dam with the tail water level.

The procedure described above is repeated until the free surface polygon $\overline{\text{ABCDEF}}$ beginning at the fixed point A ends at the fixed point F. For the problem under consideration, a reasonable fit was achieved for the following values (determined by the above procedure).

Zone No. (i)	1	2	3	4	5
J_i'	0·0134	0·1112	1·3199	2·1385	0·1790
v_i' (mm/s)	6·0	7·2	8·5	11·0	32·0

The free surface $\overline{\text{ABCDEF}}$ in Fig. 4.4 is drawn according to the values above. Using the determined values v_i' and the given values $\bar{D}_i'$ we arrive at the following orders of the Reynolds numbers (corresponding to $\nu = 10^{-6}$ m/s, i.e. to the water in $\approx$20°C).

$$\overline{\text{Re}}_1' = 600{\cdot}0,\ \overline{\text{Re}}_2' = 358{\cdot}7,\ \overline{\text{Re}}_3' = 84{\cdot}6,\ \overline{\text{Re}}_4' = 82{\cdot}5,\ \overline{\text{Re}}_5' = 1587{\cdot}5$$

and thus at the following scale relations:

$$\lambda_{n_1}{}^6 = \frac{6{\cdot}000 + \lambda_{\bar{D}}{}^{-\frac{3}{2}}}{6{\cdot}000 + 1}$$

$$\lambda_{n_2}{}^6 = \frac{3{\cdot}587 + \lambda_{\bar{D}}{}^{-\frac{3}{2}}}{3{\cdot}587 + 1}$$

$$\lambda_{n_3}{}^6 = \frac{0{\cdot}846 + \lambda_{\bar{D}}{}^{-\frac{3}{2}}}{0{\cdot}846 + 1}$$

$$\lambda_{n_4}{}^6 = \frac{0{\cdot}825 + \lambda_{\bar{D}}{}^{-\frac{3}{2}}}{0{\cdot}825 + 1}$$

$$\lambda_{n_5}{}^6 = \frac{15{\cdot}875 + \lambda_{\bar{D}}{}^{-\frac{3}{2}}}{15{\cdot}875 + 1}$$

and

$$\lambda_v{}^2 = \lambda_{\bar{D}}$$

Observe that the similarity problem under consideration can hardly be solved by assigning unity to all scales. Indeed, let us assume for a while that λ_D is unity, and thus that the Zone 1 in the model is formed by the prototype material having 'grains' as big as $\approx$10 cm. Even if we were to decide that the thickness of Zone 1 in the vicinity of the point A should consist of only two grain layers, this would imply that the model thickness of Zone 1 at A must be $\approx$200 mm. Considering that the prototype thickness of Zone 1 at A is approximately 2·5 meters (see Fig. 4.4), it follows that our model must be of the scale

$$\lambda_l = \frac{20}{250} = \frac{1}{12 \cdot 5}$$

But this implies that the model must possess a height of 100/12·5 = 8 m, and a width of 300/12·5 = 24 m—almost a real dam within the laboratory! . . . Hence, there is no alternative but to reduce the model dimensions by means of choosing $\lambda_{\bar{D}} \ll 1$. Let us select, for example

$$\lambda_{\bar{D}} = \frac{1}{10}$$

In this case, the effective size of the model granular material forming the Zone 1 will be $\bar{D}_1'' = 0 \cdot 1 \times 10 = 10$ mm. Let us assume that the thickness of the model Zone 1 at the point A is selected as 30 mm (three grain diameters). This would imply that the model scale is

$$\lambda_l = \frac{3}{250} = \frac{1}{83 \cdot 3}$$

while the model height and width are 1·20 m and 3·60 m—a large but not unreasonable model

Using the grain size scale $\lambda_{\bar{D}} = 1/10$ and the porosity scales

$$\lambda_{n_1} = \left(\frac{6 \cdot 000 + 10^{\frac{3}{2}}}{7 \cdot 000}\right)^{\frac{1}{6}} = 1 \cdot 324$$

$$\lambda_{n_2} = \left(\frac{3 \cdot 587 + 10^{\frac{3}{2}}}{4 \cdot 587}\right)^{\frac{1}{6}} = 1 \cdot 405$$

$$\lambda_{n_3} = \left(\frac{0 \cdot 846 + 10^{\frac{3}{2}}}{1 \cdot 846}\right)^{\frac{1}{6}} = 1 \cdot 613$$

$$\lambda_{n_4} = \left(\frac{0 \cdot 825 + 10^{\frac{3}{2}}}{1 \cdot 825}\right)^{\frac{1}{6}} = 1 \cdot 616$$

$$\lambda_{n_5} = \left(\frac{15 \cdot 875 + 10^{\frac{3}{2}}}{16 \cdot 875}\right)^{\frac{1}{6}} = 1 \cdot 188$$

We arrive at the following values determining the nature of five model zones

Zone No. (i)	1	2	3	4	5
$\bar{D}_i''$ (mm)	10·0	5·0	1·0	0·75	5·0
n_i''	0·27	0·24	0·27	0·29	0·30

The scale of filtration velocity being

$$\lambda_v = \sqrt{\left(\frac{1}{10}\right)} = \frac{1}{3{\cdot}16}$$

4.4 Non-steady State Filtration

If the difference in water levels generating filtration flow varies with time t, then the properties of filtration flow will also vary with t (non-steady state or non-stationary filtration). Accordingly, with respect to the examples above, the filtration phenomena must be regarded as non-stationary if in Example 4.1, say, the level of the free surface of the river, or in Example 4.2 the level of the tail water of the dam, varies with time. Clearly, in such cases, time t will be an additional characteristic parameter, while the dimensionless time (Strouhal number)

$$\mathrm{St} = \frac{vt}{l} \tag{4.35}$$

will be an additional dimensionless variable. From the identity of the model and prototype values of this dimensionless variable, we obtain the following expression for the time scale

$$\lambda_t = \frac{\lambda_l}{\lambda_v} \tag{4.36}$$

where λ_l and λ_v are model scale and filtration velocity scale, respectively. Accordingly, if, for example, the filtration through the earth dam shown in Fig. 4.4 is non-stationary, and if it is due to the variation in the tail water level, then any variation of the prototype tail water level during a time interval t' must take place in the model during the corresponding time interval $t'' = \lambda_t \times t'$, where λ_t is given by (4.36). Only if such a variation of the model tail water level is provided can the kinematic similarity of the time dependent filtration be achieved.

Considering (4.31) the time scale given by (4.36) can be expressed as follows:

$$\lambda_t = \frac{\lambda_l}{\sqrt{\lambda_{\bar{D}}}} \tag{4.37}$$

Thus, if the system in Example 4.1 were non-stationary, then the time scale would be

$$\lambda_t = \frac{1/20}{1} = \frac{1}{20} \qquad (\text{when } \lambda_{\bar{D}} = 1)$$

and

$$\lambda_t = \frac{1/20}{\sqrt{\frac{1}{2}}} = \frac{1}{14{\cdot}15} \qquad (\text{when } \lambda_{\bar{D}} = \tfrac{1}{2})$$

while in the case of Example 4.2 it would be

$$\lambda_t = \frac{1/83{\cdot}3}{\sqrt{(0{\cdot}1)}} = \frac{1}{26{\cdot}2}$$

4.5 Similarity of Filtration Flow obeying the Darcy Law

When the order of the Reynolds number $\overline{\text{Re}} = v\bar{D}/\nu$ becomes smaller than unity, then the curves plotted in Figs. 3.1 and 3.2 become practically indistinguishable from the 45° straight lines. The structure of the eqn. (4.18) is in agreement with this. Indeed, when $\overline{\text{Re}} < \approx 1$, the importance of the first term in the expression within the brackets becomes less than one percent, and thus $\bar{\Pi}_J$ can be regarded as proportional to $1/\overline{\text{Re}}$. The proportionality between $\bar{\Pi}_J$ and $1/\overline{\text{Re}}$ implies proportionality between v and J, and thus the validity of the Darcy law

$$v = kJ$$

From this expression of the Darcy law we obtain

$$\lambda_v = \lambda_k \lambda_J \tag{4.38}$$

and since $\lambda_J = 1$

$$\lambda_v = \lambda_k \tag{4.39}$$

Hence, if both model and prototype Reynolds numbers $\overline{\text{Re}}''$ and $\overline{\text{Re}}'$ are less than the order ≈ 1, then the scale of filtration velocities is equal to the ratio k''/k' of the model and prototype permeability coefficients. From

(4.13) it follows that if the model and prototype granular media were geometrically similar, then the scale of the permeability coefficient would be

$$\lambda_k = \lambda_\gamma \, \lambda_\mu^{-1} \, \lambda_D^{2} \tag{4.40}$$

which for the case of the same fluid in the model and prototype reduces into

$$\lambda_k = \lambda_D^{2} \tag{4.41}$$

However, as has been pointed out earlier, it would hardly be practical to control and rely on the geometric similarity of the porous media (and

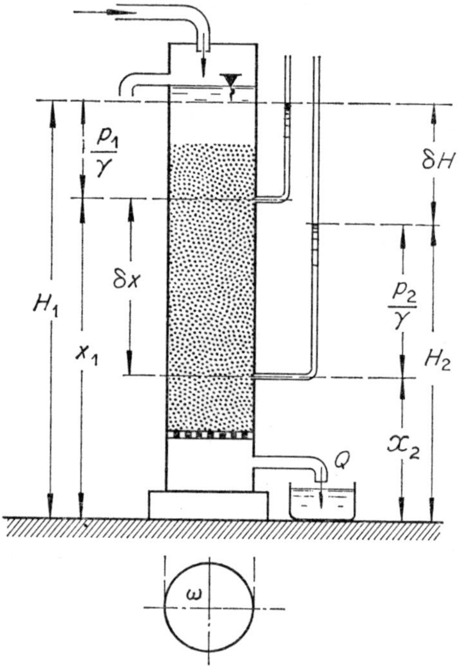

FIG. 4.5

this is precisely why the effective diameter $\bar{D}$ was introduced). In the laboratory, the permeability coefficient is usually determined by an apparatus such as that shown schematically in Fig. 4.5. Measuring $J = -\delta H/\delta x$ and $v = Q/\omega$, one computes $k = v/J$*. Having determined k' and k'', we obtain the velocity scale as the ration k''/k'.

* An extensive account on various methods of determining the values of the permeability coefficient in the field can be found, e.g. in the dissertation of K. Çeçen[7], where he presents also his own original method. (See also the Refs. 1, 2, 3, 8.)

If the medium under investigation is heterogeneous, then in order to maintain the same velocity scale for all parts (1, 2, . . . n) of the medium, the model materials, i.e. the values k_1'', k_2'', . . . k_n'' must be such as to satisfy

$$\frac{k_1''}{k_1'} = \frac{k_2''}{k_2'} = . . . \frac{k_n''}{k_n'} = \lambda_k = \text{const.} \tag{4.42}$$

In the case of a non-stationary filtration, the model and prototype values of the Strouhal number must be identical, which implies that

$$\frac{\lambda_t \lambda_v}{\lambda_l} = 1 \tag{4.43}$$

and thus

$$\lambda_t = \frac{\lambda_l}{\lambda_v} = \frac{\lambda_l}{\lambda_k} \tag{4.44}$$

must also be valid.

Hydraulic filtration models, made of granular material and operating with fluid, are cumbersome, while the precision in the determination of the streamlines, from the patterns left by the dye, is rather poor. Hence, it is advisable to adopt such models only when no other alternative is possible, that is when the Darcy law is not valid for all stages and in all parts of the medium under investigation (the examples in the preceding section). If, on the other hand, the Darcy law is valid (i.e. if $\overline{\text{Re}} < \approx 1$), then it would be far more advantageous to use electrical (or electro-analogical) models which are easier to operate and which give incomparably more precise results as far as the determination of the streamlines (flow net) is concerned. In 1940 S. A. Khristianovich[12] demonstrated that electrical models can also be used for the reproduction of *non-linear* filtration ($\text{Re} > \approx 1$). However, the utilization of electrical models for flows not obeying the Darcy law, has not so far become popular.

The principle of electrical models (which was, according to Soviet sources, first introduced by N. N. Pavlovsky (1918), and according to German sources, by P. Böss (1920)) rests on the analogy between the mathematical laws governing filtration and electrical currents summarized in Table 4.1 which is adopted from Ref. 5.

The scope of the present book does not extend to cover details of the electroanalogy method and the reader is referred to the special literature on filtration[1,2,3,5]. In principle, however, the two-dimensional model consists of a two-dimensional conductor having the shape of the porous medium under investigation. As is clear from the sketch (p. 102), the contour

Table 4.1

Filtration	Electrical current
Energy height H Filtration coefficient k	Electrical potential U Specific conductivity c
Filtration velocity $\vec{v}$ Darcy law:	Current density $\vec{i}$ Ohm's law:
$\vec{v} = -k \times \overrightarrow{\text{grad. } H}$	$\vec{i} = -c \times \overrightarrow{\text{grad. } V}$
Laplace eqn. for energy height	Laplace eqn. for electrical potential
$\frac{\partial^2 H}{\partial x^2} + \frac{\partial^2 H}{\partial y^2} + \frac{\partial^2 H}{\partial z^2} = 0$	$\frac{\partial^2 V}{\partial x^2} + \frac{\partial^2 V}{\partial y^2} + \frac{\partial^2 V}{\partial z^2} = 0$
Boundary conditions: 1. Impermeable surface	Boundary conditions: 1. Non-conductive surface
$\frac{dH}{dn} = 0$	$\frac{dV}{dn} = 0$
where n — normal to the surface 2. Equi-H — surface	where n — normal to the surface 2. Equi-V— surface (equipotential)
$H = f(x, y, z) = \text{const}$	$V = f(x, y, z) = \text{const}$
Flow rate Q	Current I
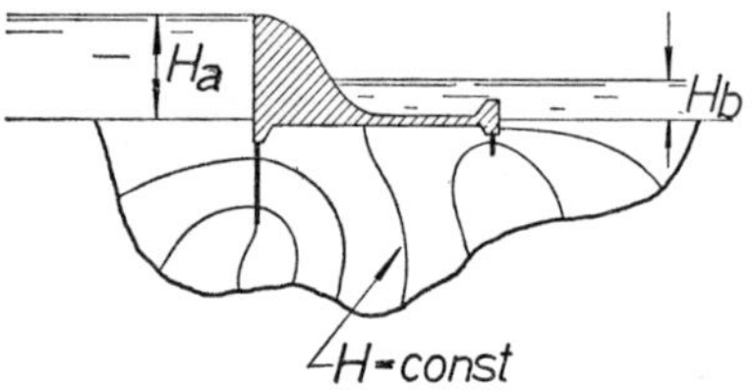	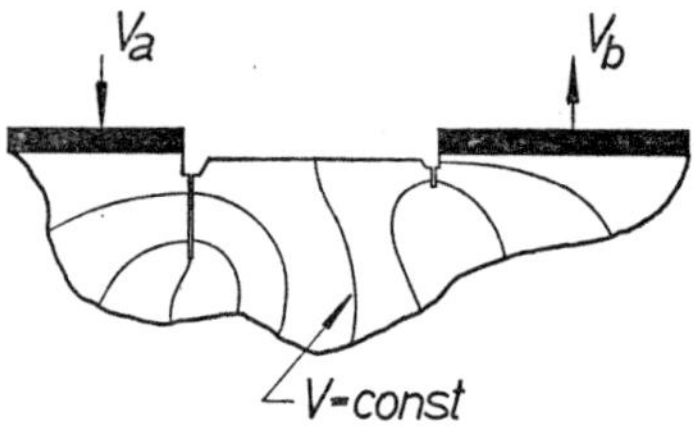

of the conductor coincides with the boundary between the permeable granular medium and impermeable ground, while the inlet and outlet of the filtration flow are substituted by brass or copper electrodes (black stripes in the sketch on the right). From the structure of the Laplace equation, it is clear that the geometry of the flow net (consisting of orthogonal families of the streamlines and equipotential lines) does not depend on the filtration coefficient k or, with respect to the electrical equivalent, on the conductivity c. Thus, the absolute value of the conductivity c of the conductor used is not relevant. Accordingly, the two-dimensional conductor can be formed by a variety of materials.

Tin foil sheets (0·01 to 0·02 mm thick), various electrolytes (solutions of $CuSO_4$, NaCl, Na_2CO_3 etc.), wood or cardboard surfaces painted with varnish, containing conductive graphite powder, are among successfully used conductors. In the Soviet Union the use of mass produced electroconductive paper became popular in the last two decades.

If the medium is heterogeneous, then, in accordance with (4.42), the proportionality

$$\frac{c_1}{k_1'} = \frac{c_2}{k_2'} = \ldots \frac{c_n}{c_n'} \tag{4.45}$$

must be valid even if the absolute values of c_i are still of no importance. The validity (4.45) is achieved by an appropriate (proportional) variation in the concentration of the agent producing conductivity, by adjusting the thickness of the pack of tin foil sheets by varying the amount of the dissolved $CuSO_4$, etc. In the case of an electrolyte, the regions of different permeability (conductivity) must be separated by an impermeable boundary. The use of electrolyte can be especially effective if the permeability coefficient is (or can be treated as) a *continuous* function of position in porous medium

$$k' = f(x', y')$$

i.e.

$$k' = k_o'\varphi(\xi, \eta)$$

where k_0 is the value of the permeability coefficient that is selected as 'typical'. In such cases, all that is required to give an appropriate topography to the wax (paraffin) forming the bottom of the container. Indeed, since the conductivity c is proportional to the thickness δ of the conductor, the required proportionality

$$\delta \sim c \sim k \sim \varphi(\xi, \eta) \tag{4.46}$$

will automatically be satisfied. If k varies as a linear function of the position ξ, η in porous medium (or if its actual variation can be approximated by a linear function), then it is sufficient to tilt the flat glass bottom of the container by an appropriate angle α as implied by Fig. 4.6.[14]

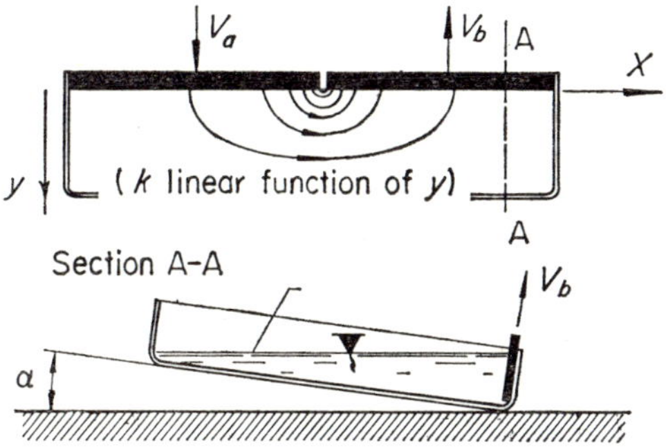

FIG. 4.6

The determination of the flow net on an electroanalogical model rests on finding the points of equal potential, by a probe, and consequently by revealing the equipotential lines. Subsequently, one obtains the streamlines as a family of the curves that are orthogonal to the determined family of the equipotential lines.

The electrical circuit of the electroanalogical model is, in principle, a Wheatstone bridge (Fig. 4.7a and b). The elements between the points

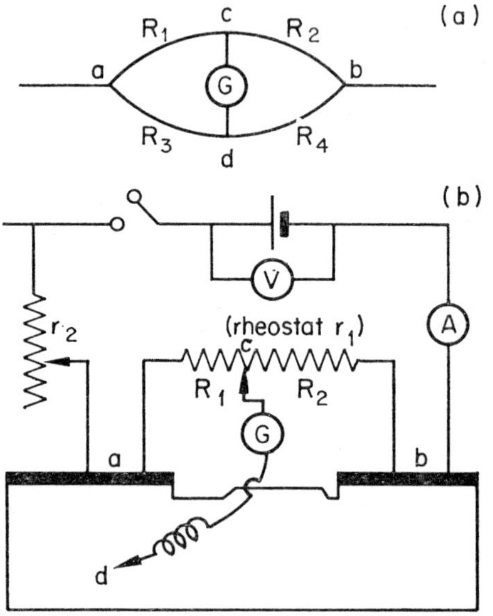

FIG. 4.7a, b

a, b, c and d of the circuit in Fig. 4.7b correspond to the same points of the scheme in Fig. 4.7a. The probe *d* is moved on the surface of the conductor (model) until the galvanometer G, on the 'bridge' indicates zero, i.e. until the following identity among the resistances is established

$$\frac{R_1}{R_3} = \frac{R_2}{R_4} \tag{4.47}$$

When the relation above is valid, then the probe *d* is on one of the points of the equipotential line corresponding to the specified fraction of the total potential (head) difference $V_a - V_b$. A certain fraction of $V_a - V_b$ is provided by a certain position of the point c on the rheostat r_1.

When all points forming the equipotential line corresponding to a particular fraction of $V_a - V_b$ are revealed, the position of the point c on the rheostat is altered to provide a new fraction of the total potential difference, and one sets out to determine the points of the new equipotential line. The value of the total potential difference $V_a - V_b$ is read from the voltmeter V, the current of the system being indicated by the ammeter A (and being adjusted by the rheostat r_2).*

If the prototype filtration flow has a free surface (Fig. 4.8), then the boundary $a''b''$ of the conducting medium in the model must be geometrically similar to the prototype free surface. Since, however, the initial shape

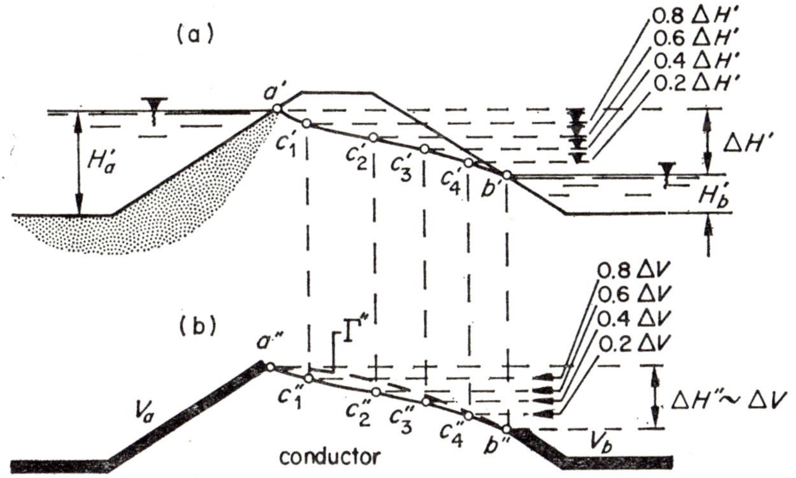

FIG. 4.8a and 4.8b

of the free surface is not known, its counterpart in the model is determined by successive approximations. At the free surface the manometric pressure is zero, and thus the head H corresponding to any point of the free surface is simply equal to the level y of the point, i.e.

$$H = y \tag{4.48}$$

is valid for any point of the free surface. This implies that if the prototype head difference $\Delta H' = H_a' - H_b'$ is divided in, say, five equal regions (Fig. 4.8a), then the points C_1', C_2', C_3' and C_4' of the free surface must correspond to the values of the head $0{\cdot}8\Delta H'$, $0{\cdot}6\Delta H'$, $0{\cdot}4\Delta H'$ and $0{\cdot}2\Delta H'$, respectively. Similarly, the model points C_1'', C_2'', C_3'' and C_4'' must correspond to the potentials $0{\cdot}8\Delta V$, $0{\cdot}6\Delta V$, $0{\cdot}4\Delta V$ and $0{\cdot}2\Delta V$ (where $\Delta V = V_a - V_b$). Hence, the following procedure can be suggested. Estimate a free surface curve and cut the conductor along a curve Γ'' that is higher than the expected position of the free surface (Fig. 4.8b). Subsequently, divide the model distance $\Delta H''$ into five equal regions and

* A variety of circuit schemes used for electrical model can be found[5].

draw parallel lines separating these regions. Adjust the rheostat so as to give $0{\cdot}8\Delta V$ and move the probe along the uppermost line corresponding to that voltage until the galvanometer indicates zero, that is, until the point C_1'' is found. In a similar manner, find the points C_2'', C_3'' and C_4''. Then cut the conductor along a curve that is *between* the curve Γ'' and the curve determined by the sequence of the points C_i''. Repeat the procedure and find a new sequence of the points C_i'' (second approximation). Continuing in this manner, one will finally arrive at a sequence of such points

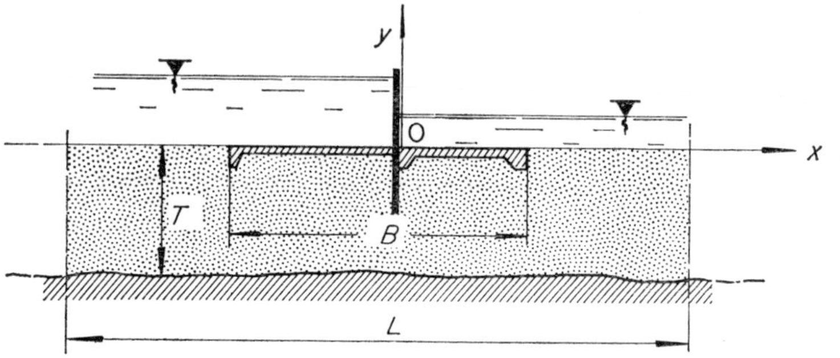

FIG. 4.9

C_i'' that are practically on the boundary of the conductor. These points C_i'' (or the boundary of the conductor) imply the free surface.

In some cases the permeable layer is of a limited thickness in the direction $-y$, but it is unlimited in the direction $\pm x$ (Fig. 4.9). In such cases, it is sufficient to reproduce in the model the length

$$L \approx B + (3 \text{ to } 4)T \tag{4.49}$$

only. If the permeable layer is unlimited in both directions $-y$ and $\pm x$ (Fig. 4.10), then one can say that the filtration taking place outside the radius

$$R \approx 1{\cdot}5B \quad \text{or} \quad R \approx 3S \text{ (whichever is larger)} \tag{4.50}$$

is negligible. Accordingly, the size of the rectangular conductor plate can be selected as $R \times 2R$.*

The scope of the present book does not permit the inclusion of details of electrical models of filtration flows. For these, the reader is recommended to consult Chapter 12 of Ref. 2. In this excellent special treatise on filtration, in addition to the electrical analogy method which was outlined above, and which can be referred to as a *continuous method*, the *discrete*

* Clearly, the same numerical orders apply to the size of the filtration models that are not necessarily electrical models[5].

(*or discontinuous*) electrical analogy method[15,16] is also represented. The latter method reveals the potential field by supplying values of the potential for certain discrete points in the medium.

Before closing the chapter, it should also be mentioned that hydraulic and electrical models are not the only means of reproducing filtration flows. Any device where the potential flow can be achieved can be used for the study of linear filtration (obeying the Darcy law). Thus, the model study of linear filtration can be carried out, and is often carried out (since R. Dachler[17]) by using the *Hele-Shaw* apparatus—a popular device for the

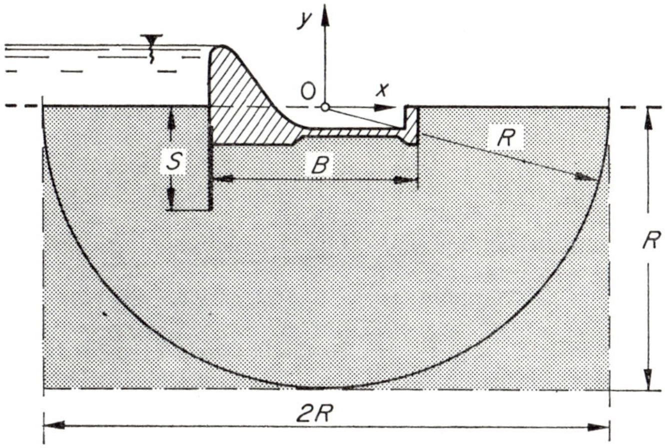

FIG. 4.10

study of potential flows.* By adjusting the thickness of the plates placed between the parallel glass walls, even heterogeneous conditions can be reproduced in a Hele-Shaw model. Similarly, potential flow, and thus the filtration, can be studied by the *ion-motion analogy*[19,20,21] and the *membrane analogy*[15,22,23], methods. The description of these methods can also be found in Ref. 2.

REFERENCES

1. P. Ya. Poluborinova-Kochina, *Theory of Ground Water Movement*, Princeton University Press (1962).
2. J. Bear, D. Zaslavsky, S. Irmay, *Physical Principles of Water Percolation and Seepage*, UNESCO, Paris (1968).
3. M. Muskat, *The Flow of Homogeneous Fluids through Porous Media*, J. W. Edwards, Inc., Michigan (1946).

* A systematic analysis of the scales for Hele-Shaw filtration models is due to J. Bear[18].

4. A. Verruijt, *The Theory of Groundwater Flow*, Macmillan, London (1970).
5. V. I. Aravin, S. N. Numerov, *Filtration in Hydraulic Structures* (In Russian), Gostekhizdat, Moscow (1958).
6. G. K. Batchelor, *An Introduction to Fluid Dynamics*, Cambridge University Press (1967).
7. K. Çeçen, *Die Ermittlung des Durchlässigkeitsbeiwertes im Zusammenhang mit bautechnischen Bodenuntersuchungen*, Verlag Wasser und Boden (1967).
8. I. I. Agroskin, G. T. Dmitriyev, A. N. Pikalov, *Hydraulics*, Gosenergoizdat, Moscow (1944).
9. E. Lindquist, *The Flow of Water through Porous Soil*, Ier Congr., Grand Barrages, Stockholm (1933).
10. G. Schneebeli, 'Experiments on the Range of Validity of Darcy's Law and the Appearance of Turbulence in a Filtrating Flow'. *La Houille Blanche*, Mars-Avril, No. **2** (1955).
11. P. Nemenyi, *Wasserkraft und Wasserwirtschaft*, **29**, 157 (1934).
12. S. A. Khristianovich, 'Flow of Ground Water, not obeying Darcy's Law', *Applied Mathematics and Mechanics*, **IV**, No. 1, Moscow (1940).
13. G. Cohen de Lara, Head Loss in a Porous Medium Based on the Hydrodynamic Equilibrium of the Mass', *La Houille Blanche*, Mars-Avril, No. **2**, 1955.
14. M. S. Yalin, 'Über die Sickerströmung in einem stetig heterogenen Raum', *Bulletin of Techn. University of Istanbul*, Vol **7**. (1954).
15. W. J. Karplus, '*Analog Simulation Solution of Field Problems*', McGraw-Hill (1958).
16. G. De Josselin De Jong, 'Electric Analogue Models for Solution of Geohydrological Problems', *Water*, Vol. **46**, No. 4 (in Dutch) (1962).
17 R. Dachler, *Grundwasserströmung*, Springer Verlag (1936).
18. S. Bear, 'Scales of Viscous Analog Models for Ground Water Studies', *Proc. A.S.C.E.*, Hydr. Div., **86**, HT2 (1960).
19. R. D. Wyckoff, H. G. Botset, 'An Experimental Study of Motion of Particles in Systems of Complex Potential', *Physics*, **5** (1934).
20. M. Muskat, *Physical Principles of Oil Production*, McGraw-Hill (1949).
21. H. G. Botset, 'The Electrolytic Model and its Application to the Study of Recovery Problems', *Trans. A.I.M.E.*, Petrol Div., **165** (1946).
22. V. E. Hansen, 'Complicated Well Problems Solved by the Membrane Analogy', *Trans. Amer. Geophys. Un.*, **33** (1952).
23. G. De Josselin De Jong, 'Singularity Distributions for the Analysis of Multiple Fluid Flow through Porous Media', *J. Geophys. Res.*, **65** (1960).

5 Unidirectional Flows with a Free Surface, River and Open Channel Models

5.1 General

Model tests are usually carried out if the properties involved in the design of a hydraulic project cannot be determined theoretically, or if such a determination cannot be regarded as reliable. Accordingly, with the exception of some investigations mainly for scientific rather than technical purposes, one cannot expect the subject of a model test to be, say, a uniform flow in a regular prismatic channel. Consequently, one must be prepared to face the presence of irregular flow boundaries, and of a complicated, often three-dimensional, structure of the flow. Considering this, scale relations are developed in the following study for the most general case of river and channel flow—non-uniform flow in a non-prismatic channel. The formulation of simpler (special) cases, such as prismatic channels, uniform flows, etc. is then automatically included in the general forms. First, steady state or stationary flows are considered and subsequently the method is generalized to non-steady flows. Since almost all the flows in civil engineering practice are turbulent, the consideration of laminar flows is omitted. In the present chapter, the presence of a *rigid bed* is assumed, the consideration of a *mobile bed* (consisting of erodible granular material) being given in Chapter 6. The term rigid bed should not necessarily be understood in the literal sense of the word rigid. If the flow bed consists of loose granular material but the stage of the flow is below the critical stage, corresponding to the initiation of sediment transport, and the granular material is thus not in motion, then for such stages the movable bed is acting as a rigid bed. Similarly, if sediment is being transported in the vicinity of the bed (such transported material is termed *bed load*) and if the flow stages of the model tests do not differ appreciably from each other, so that there is negligible change in the surface geometry of the bed, then although granular material is moving the flows can still be treated as if taking place over a rigid bed. Such a 'rigid' bed will possess a particular wave-like geometry and possess a certain value of effective sand roughness (k_s) equivalent to the geometry of sand waves.

5.2 Dynamic Similarity of a Non-uniform Steady State Flow in a Non-prismatic Channel

5.2.1 General Case In Example 1.4 it was shown that the stationary non-uniform flow in a non-prismatic channel, possessing a specified geometry and roughness, can be defined by the following set of characteristic parameters (see the set (1.66))

$$\mu, \rho, R_o, k_s, v_o, S, g \tag{5.1}$$

and thus by the following dimensionless variables

$$X_1 = \frac{v_o R_o}{\nu}; \; X_2 = \frac{k_s}{R_o}; \; X_3 = S; \; X_4 = \frac{v_o^2}{g R_o} \tag{5.2}$$

where R_o and v_o are the values corresponding to a section O—O, selected as typical. Hence, if geometric similarity is provided, then dynamic similarity of the non-uniform flow under consideration is given by

$$\lambda_{X_1} = 1; \; \lambda_{X_2} = 1; \; \lambda_{X_3} = 1; \; \lambda_{X_4} = 1 \tag{5.3}$$

i.e. by

$$\left.\begin{aligned} \lambda_U \, \lambda_l \, \lambda_\rho \, \lambda_\mu^{-1} &= 1 \\ \lambda_{ks} \, \lambda_l^{-1} &= 1 \\ \lambda_S &= 1 \\ \lambda_U^2 \, \lambda_g^{-1} \, \lambda_l^{-1} &= 1 \end{aligned}\right\} \tag{5.4}$$

Clearly, the scale relations which follow directly from (5.2) contain the scales λ_{Ro} and λ_{vo} (rather than λ_l and λ_U). But since, in dynamically similar systems, the scales of *all* lengths and *all* velocities are equal, the scales λ_{Ro} and λ_{vo} are replaced by the geometric model scale λ_l and by the general velocity scale λ_U. Similarly, in the following we will use R and v rather than R_o and v_o. For the sake of simplicity in the present considerations, we will not introduce the dimensionless co-ordinates ξ, η, ζ and their scales λ_ξ, λ_η, λ_ζ. This, however, implies automatically that any comparison between the model and prototype should be made solely for the corresponding points (or sections), i.e. satisfying the conditions $\lambda_\xi = 1$; $\lambda_\eta = 1$; $\lambda_\zeta = 1$ (or only $\lambda_\xi = 1$).

When designing a dynamically similar model, one can select freely the scales of three dimensionally independent parameters; the rest of $n - 3$ scales (in our case $7 - 3 = 4$ scales) being determined by the similarity criteria (in our case by the eqns. (5.4)). Since the acceleration due to gravity is one of the characteristic parameters, we have

$$\lambda_g = 1 \tag{5.5}$$

On the other hand, since in conventional models the prototype fluid (water) is used

$$\lambda_\mu = 1 \quad \text{and} \quad \lambda_\rho = 1 \tag{5.6}$$

But, if the scales of three dimensionally independent parameters are equal to unity, then all scales are equal to unity. Hence, the validity of (5.5) and (5.6) implies that the realization of a dynamically similar small scale model ($\lambda_l < 1$), is in general, impossible.

5.2.2 Fully Developed Turbulent Flow If the flow in both model and prototype is rough turbulent, i.e. if

$$\approx 70 < \frac{v_*'' k_s''}{\nu} < \frac{v_*' k_s'}{\nu} \tag{5.7}$$

is valid for all parts and stages of the non-uniform flow under investigation, then the parameter μ can be excluded from consideration. In this case we have only two scales that are equal to unity

$$\lambda_g = 1; \; \lambda_\rho = 1 \tag{5.8}$$

Hence, one scale, e.g. the model scale λ_l, can be selected freely, and thus the realization of a small scale model ($\lambda_l < 1$) becomes possible. The scales of the remaining characteristic parameters being determined by the following reduced version of (5.4)

$$\left.\begin{aligned} \lambda_{k_s} &= \lambda_l \\ \lambda_S &= 1 \\ \lambda_U &= \sqrt{(\lambda_l)} \end{aligned}\right\} \tag{5.9}$$

while the scale of a property A is given by

$$\lambda_A = \lambda_\rho^{-x_A} \, \lambda_l^{-y_A} \, \lambda_U^{-z_A} \tag{5.10}$$

From (5.9) it is clear that the model under consideration possesses the following properties:

(i) The roughness is scaled down in the same proportion as the external dimensions of the flow (the model is geometrically similar, even with respect to its roughness).
(ii) The bed slope S is identical to that of the prototype. Accordingly, the slope of the free surface at any section of the model is also identical to that of the corresponding section in the prototype.
(iii) The velocity scale is equal to the square root of the model scale. The model having this property is referred to as a *Froudian model.* Observe that the third eqn. in (5.9) is due to the Froude number $X_4 = v^2/gR$.

Since μ is no longer a characteristic parameter, and thus since the Reynolds number $X_1 = vR/\nu$ is no longer a dimensionless variable, the friction factor c is a function of the relative roughness $X_2 = k_s/R$ only, i.e.

$$\frac{v}{v_*} = c = \varphi_c\left(\frac{k_s}{R}\right) = \varphi_c(X_2) \tag{5.11}$$

is valid. Since the model and prototype are geometrically similar, they possess, in all corresponding regions, the same form of the function φ_c. Furthermore, $\lambda_{X_2} = 1$, that is, the model and prototype values of X_2, are the same. Hence, the value of the function $\varphi_c(X_2)$ is the same in model and prototype, and thus

$$\lambda_c = 1 \quad \text{or} \quad \lambda_v = \lambda_{v_*} \tag{5.12}$$

is valid, for all corresponding regions and stages. Since the model is Froudian, $\lambda_v = \sqrt{(\lambda_l)}$, and since it is a small scale model, $\lambda_l < 1$, we have $\lambda_v < 1$, i.e. $\lambda_{v_*} < 1$, and thus $v_*'' < v_*'$. On the other hand $k_s'' < k'$ is also valid (for $\lambda_{k_s} = \lambda_l$ while $\lambda_l < 1$). Hence, in a small scale Froudian model, the inequality

$$\frac{v_*''k_s''}{\nu} < \frac{v_*'k_s'}{\nu} \tag{5.13}$$

is always satisfied, and consequently the fulfilment of the condition (5.7) depends entirely on the fulfilment of

$$\approx 70 < \frac{v_*''k_s''}{\nu} \tag{5.14}$$

alone. Considering that $\lambda_\nu = 1$, while $\lambda_{v_*} = \lambda_v$, the inequality (5.14) can be expressed as

$$\lambda_v \lambda_{k_s} > \approx \frac{70}{\dfrac{v_*' k_s'}{\nu}} \tag{5.15}$$

or as

$$\lambda_l > \approx \left(\frac{70}{\dfrac{v_*' k_s'}{\nu}} \right)^{\frac{2}{3}} \tag{5.16}$$

since $\lambda_v = \sqrt{(\lambda_l)}$ while $\lambda_{k_s} = \lambda_l$. It follows that the model scale cannot be selected less than a certain lower limit determined by the prototype value of the Reynolds number $v_*' k_s' / \nu$.

Let us see how the condition (5.16) compares with various types of rivers. Consider (as the prototype) a river in a mountainous region which streams over a bed formed by boulders with relatively high velocity. Such a river usually has a rather steep slope, while its flow depth is moderate. The following characteristics of the North Saskatchewan River, are typical for this river during intermediate flow rates (near Drayton Valley, Alberta, Canada)[1]

$$h' = 2{\cdot}50\,\text{m};\; S' = 0{\cdot}0015;\; k_s' = 50\,\text{mm} \tag{5.17}$$
$$(g = 9{\cdot}81\ \text{m/s};\; \nu = 10^{-6}\ \text{m}^2/\text{s}\ (\approx 20°\text{C}))$$

Considering that in the midst of a natural river the flow can almost invariably be treated as two-dimensional, we can adopt the simpler relations of a two-dimensional flow.

In the case of a two-dimensional, non-uniform flow, the value of the shear stress τ_0 acting on the bed is given by

$$\tau_0 = \gamma J h \tag{5.18}$$

while the value of the shear velocity is

$$v_* = \sqrt{\left(\frac{\tau_0}{\rho}\right)} = \sqrt{(gJh)} \tag{5.19}$$

where J stands for the *energy gradient* $-\mathrm{d}H/\mathrm{d}x$. One can, of course, always disregard the two-dimensional treatment and use the expressions $\tau_0 = \gamma RJ$ and $v_* = \sqrt{(gRJ)}$ (where $R = \omega/\chi$ hydraulic radius) rather than those given by (5.18) and (5.19). However, for the purposes of the present study, this would be an unnecessary elaboration. If, as in the present example, the river is considered as a whole, that is, if the attention is not focused on a particular region of the river where, for a given flow rate Q, the gradient J is fixed, then we have no alternative but to regard J, and thus τ_0 and v_* as quantities fluctuating (with the distance x) about certain average values. By assigning a value to the flow depth (such as $h' = 2{\cdot}50$ m) we imply that the average value of the fluctuating quantity J is the average slope S of the river bed, while the average values of τ_0 and v_* (along x) are

$$\gamma hS \quad \text{and} \quad \sqrt{(ghS)}$$

respectively. It follows that, if the river is considered as a whole, then we have no alternative but to treat the non-uniform flow as uniform. Accordingly, the average value of v_* corresponding to the prototype as defined by eqn. (5.17), should be given by

$$v_*' = \sqrt{(gh'S')} = \sqrt{(9{\cdot}81 \times 2{\cdot}50 \times 0{\cdot}0015)} = 0{\cdot}192 \text{ m/s.}$$

Hence,

$$\frac{v_*'k_s'}{\nu} = \frac{0{\cdot}192 \times 0{\cdot}05}{10^{-6}} = 9600$$

and consequently,

$$\lambda_l > \approx \left(\frac{70}{9600}\right)^{\frac{2}{3}} = \frac{1}{26{\cdot}6}$$

The river model would be perfectly reasonable if the model flow were, say, $h'' = 0{\cdot}125$ m, and thus if the model scale were

$$\lambda_l = \frac{0{\cdot}125}{2{\cdot}50} = \frac{1}{20}$$

Since $\lambda_l = 1/20$ satisfies the inequality above with a considerable margin, the condition (5.16) does not present any difficulty with respect to the prototype considered above. The situation will be quite different, however, if the prototype were a typical lowland river possessing an appreciable depth and a relatively small slope, and especially if the size k_s' of the 'equivalent sand roughness' corresponding to the irregularities of its bed surface were relatively small. Such a prototype might be characterized, for example, by

$$h' = 5{\cdot}00 \text{ m}; \; S' = 0{\cdot}0001; \; k_s' = 0{\cdot}02 \text{ m}$$

Using these values we determine

$v_*' = \sqrt{(9{\cdot}81 \times 5 \times 0{\cdot}0001)} = 0{\cdot}07$ m/s

$$\frac{v_*' k_s'}{\nu} = \frac{0{\cdot}07 \times 0{\cdot}02}{10^{-6}} = 1400$$

and consequently,

$$\lambda_l > \approx \left(\frac{70}{1400}\right)^{\frac{2}{3}} = \frac{1}{7{\cdot}38}$$

Observe that this condition can hardly be satisfied by a conventional model. Indeed, even if the model scale were exactly equal to this lower limit, the model flow depth would be as large as

$$h'' = \frac{5{\cdot}00}{7{\cdot}38} \approx 0{\cdot}68 \text{ m}$$

Imagine what width B'' such a model will have, considering that, in the case of a lowland river the width/depth ratio B'/h' is usually larger than 50/1.

Hence, the method suggested above for the design of a dynamically similar model can be applied only in some cases (large values of k_s and S); it cannot cover all cases that one encounters in engineering practice. Accordingly, other methods must be sought for the design of small scale river and channel models.

5.3 Distorted River and Channel Models

5.3.1 The Idea and Definition of Distortion Consider an open channel of wide trapezoidal cross-section as shown in Fig. 5.1. The flow in the central region B_c of such a channel (shaded area in Fig. 5.1) is practically indistinguishable from two-dimensional flow, i.e. from the flow corresponding to

$$\frac{B}{h} \to \infty \tag{5.20}$$

From the work of G. Keulegan[2] it follows that if the roughness is distributed along the wetted perimeter uniformly, and if the channel flow is in rough turbulent regime, then the width B_c of the central region can be given by the following approximate relationship

$$B_c \approx B - 2 \times 2{\cdot}5h = B - 5h \tag{5.21}$$

From (5.21) it is clear that the central region B_c can exist only if

$$\frac{B}{h} > \approx 5 \tag{5.22}$$

If the roughness is not distributed along the wetted perimeter uniformly, and if the flow is not rough turbulent, then the numerical value which appears in the relations above as 5 will, of course, be different. In the present state of knowledge it is not known exactly how this value varies, depending on the character of distribution of roughness along the wetted perimeter. Nor is it known how it varies with the Reynolds number $v_* k_s/\nu$ when it is smaller than ≈ 70, i.e. when the flow is no longer in a

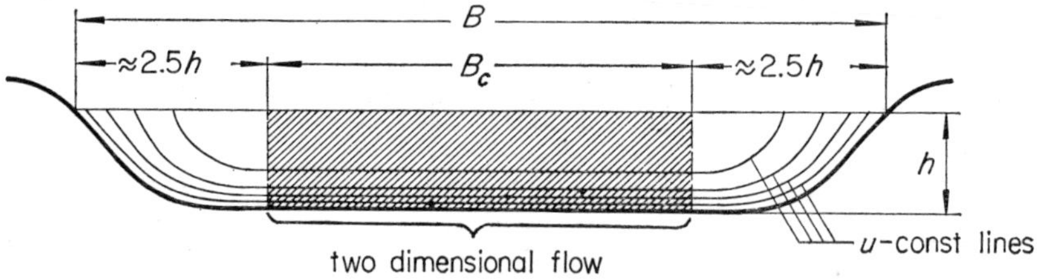

FIG. 5.1

rough turbulent regime. On the other hand, it is well-known that the size B_c of the central region can only increase, or that the value 5 can only decrease if the value of the ratio

$$\frac{(k_s)_{\text{bed}}}{(k_s)_{\text{bank}}} \tag{5.23}$$

increases (from unity onwards). In natural rivers and channels, the bed is almost always rougher than the banks, and thus by adopting the (approximate) value ≈ 5, i.e. by adopting (5.21) and (5.22) as they stand, one is on the safe side. In fact, the experiments of H. Basin[3], mentioned in the work of G. Keulegan, correspond to rectangular rather than trapezoidal sections. But here again one is on the safe side because the constant velocity lines in a trapezoidal cross-section have smaller curvature (at the bed-bank corners of the cross-section), and can thus acquire a rectilinear form (in the shaded region) within a transitional distance that is actually smaller than $\approx 2{\cdot}5h$. In this context, it should also be noted that, according to Ven Te Chow[4], in the case of a rectangular section the condition (5.22) should be replaced by a more severe, but also more vague inequality $B/h > (5 \text{ to } 10)$.

Since the flow in the central region B_c does not depend on the width B, and thus on the ratio B/h, the properties in the central region B_c'' of the

3 Undistorted model of the Van der Kloof Dam (Orange River)

4 Polystyrene dunes in a 2-ft wide flume, Hydraulics Research Station, Wallingford
(Polystyrene $D = 1{\cdot}35$ mm)

model flow cannot be affected if the model ratio B''/h'' is not equal to the prototype ratio B'/h' (Fig. 5.2). But the possibility of having inequality

$$\frac{B''}{h''} \neq \frac{B'}{h'}$$

implies that the validity of

$$\lambda_B = \frac{B''}{B'} \neq \frac{h''}{h'} = \lambda_h \tag{5.24}$$

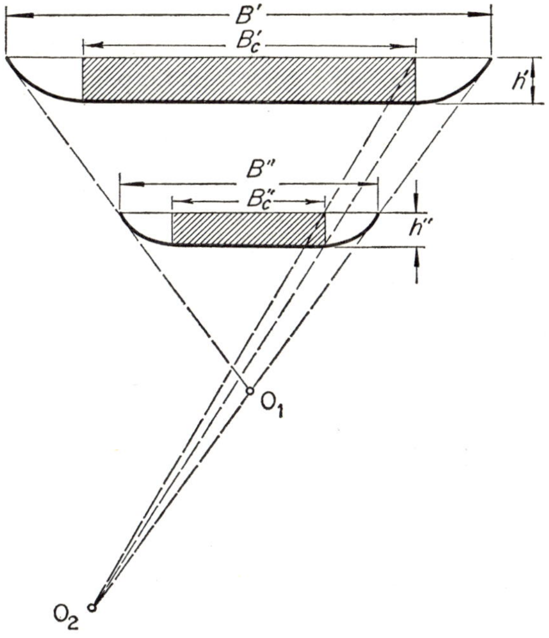

FIG. 5.2

is permissible, or that the model might possess *different* horizontal and vertical model scales λ_x and λ_y, respectively. Having λ_x smaller than λ_y, i.e. having

$$\frac{\lambda_y}{\lambda_x} = n > 1 \tag{5.25}$$

one achieves an n fold economy with respect to the model area (in comparison to a single scale model). A model corresponding to $n > 1$ is referred to as a *distorted model* as distinct from a geometrically similar or *undistorted* single scale model (corresponding to $n = 1$; the ratio n being referred to as the *distortion ratio* or simply as the *distortion.*

5.3.2 Gradually Varying River and Channel Flows Having defined the meaning of distortion, we can proceed to the determination of scales of distorted models, which are widely used in the design of river works. In the majority of cases encountered in hydraulic engineering one is concerned with the properties of sections, rather than of points of the sections. For example, one is interested in the values of average or maximum velocities corresponding to various sections, with the variation of the slope of the free surface, from one section to another; with the value of the average shear stress τ_0 acting on the wetted perimeter of a section and so on. Very often the section properties, which differ from the point properties because they vary only with the coordinate x (or with the dimensionless coordinate ξ), can be determined by means of some other section properties (usually corresponding to the same section, i.e. to the same value x or ξ). This means that in many cases the section properties can be regarded as a certain group on their own and one can attempt to design a model where at least the members of this group are reproduced in accordance with the laws of dynamic similarity. Consider, for example, the differential equation of a *gradually varying* (along x) non-uniform flow in a non-prismatic channel. This may be written for the prototype, as

$$\frac{1}{S'}\frac{\mathrm{d}h'}{\mathrm{d}x'} = \frac{1 - \dfrac{\mathrm{Fr}'}{c'^2 S'}\left(1 - \alpha \dfrac{c'^2}{\chi'}\dfrac{\partial \omega'}{\partial x'}\right)}{1 - \alpha \dfrac{B'}{\chi'}\mathrm{Fr}'} \tag{5.26}$$

This relationship, is simply the equation of motion expressed in a convenient form for practical purposes. It consists of the following section properties (corresponding to a particular section (x))

$S' = X_3'$ slope of the channel bed
h' = flow depth
ω' = cross-section area
B' = the width of the free surface
χ' = wetted perimeter

$$\mathrm{Fr} = \frac{v'^2}{gR'} = X_4' \text{ Froude number}$$

$$c' = \varphi_c\left(\frac{v'R'}{\nu}; \frac{k_s'}{R'}\right) = \varphi_c(X_1'; X_2') \text{ Friction factor}$$

and thus it gives the possibility of determining one section property by means of the other section properties. The model version of eqn. (5.26) is

$$\frac{1}{S''}\frac{\mathrm{d}h''}{\mathrm{d}x''} = \frac{1 - \dfrac{\mathrm{Fr}''}{c''^2 S''}\left(1 - \alpha \dfrac{c''^2}{\chi''}\dfrac{\partial \omega''}{\partial x''}\right)}{1 - \alpha \dfrac{B''}{\chi''}\mathrm{Fr}''} \tag{5.27}$$

Substituting each a'' by $\lambda_a a'$, the model version above can be expressed in the following manner

$$\frac{1}{S'}\frac{\mathrm{d}h'}{\mathrm{d}x'}\left[\frac{\lambda_h}{\lambda_S \lambda_x}\right] = \frac{1 - \dfrac{\mathrm{Fr}'}{c'^2 S'}\left[\dfrac{\lambda_{\mathrm{Fr}}}{\lambda_c^2 \lambda_S}\right]\left(1 - \alpha \dfrac{c'^2}{\chi'}\dfrac{\partial \omega'}{\partial x'}\left[\dfrac{\lambda_c^2 \lambda_\omega}{\lambda_\chi \lambda_x}\right]\right)}{1 - \alpha \dfrac{B'}{\chi'}\mathrm{Fr}'\left[\dfrac{\lambda_B}{\lambda_\chi}\lambda_{\mathrm{Fr}}\right]} \tag{5.28}$$

If the model is expected to be dynamically similar to the prototype only with respect to the section properties, then the prototype and model versions (5.26) and (5.27) of the dimensionless equation of motion (consisting of the section properties alone) must necessarily be identical. From comparison of (5.26) and (5.28) it follows that these equations can be identical only if each of the expressions in the square brackets (of (5.28)) is equal to unity; i.e. if the conditions

$$\frac{\lambda_h}{\lambda_S \lambda_x} = 1, \quad \frac{\lambda_{\mathrm{Fr}}}{\lambda_c^2 \lambda_S} = 1, \quad \frac{\lambda_c^2 \lambda_\omega}{\lambda_\chi \lambda_x} = 1, \quad \frac{\lambda_B}{\lambda_\chi}\lambda_{\mathrm{Fr}} = 1 \tag{5.29}$$

are satisfied. We assume that the flow cross-section is sufficiently wide so that it is possible to express the scale of the wetted perimeter χ in terms of the horizontal scale alone. Accordingly, we arrive at the following set of scales of the geometric properties of the flow cross-section

$$\lambda_h = \lambda_y, \; \lambda_\omega = \lambda_x \lambda_y, \; \lambda_B = \lambda_x, \; \lambda_\chi \approx \lambda_x \tag{5.30}$$

Substituting (5.30) in (5.29) we obtain

$$\lambda_S = \frac{\lambda_y}{\lambda_x}, \; \lambda_{\mathrm{Fr}} = \lambda_c^2 \lambda_S, \; \lambda_c^2 = \frac{\lambda_x}{\lambda_y}, \; \lambda_{\mathrm{Fr}} = 1 \tag{5.31}$$

i.e.

$$\lambda_S = \frac{\lambda_y}{\lambda_x}, \; \lambda_c^2 = \frac{\lambda_x}{\lambda_y}, \; \lambda_{\mathrm{Fr}} = 1 \tag{5.32}$$

The set (5.32) differs considerably from the set (5.3) which has been introduced at the beginning of the chapter for the design of an undistorted dynamically similar model, and which can be expressed as

$$\lambda_S = 1, \; \lambda_{\mathrm{Re}} = 1, \; \lambda_{k^s/R} = 1, \; \lambda_{\mathrm{Fr}} = 1 \tag{5.33}$$

because:

(i) the number of conditions is *reduced* by one, while

(ii) the number of scales which must satisfy these conditions is *increased* by one (instead of one model scale λ_l we have two model scales λ_x and λ_y).

The reduction in number of conditions is due to the fact that in the case of the distorted model, which is designed only for the section properties, the consideration of Re and k_s/R is substituted by the function c. This is equivalent to the assumption that the viscosity and roughness can affect a section property only by means of the friction factor c*. The increase in the number of scales, and, at the same time, the reduction of conditions provides a certain freedom in the selection of scales, and thus the possibility of realization of a small scale model. The methods so far introduced for the design of river and channel models which operate with the prototype fluid (water), that is, which must necessarily satisfy

$$\lambda_g = 1; \; \lambda_\mu = 1; \; \lambda_\rho = 1$$

can be summarized in the following manner:

Table 5.1

Type of model		Number of scales to determine	Number of conditions to satisfy	Number of scales which may be chosen
Dynamically similar model	general	4 viz: λ_l, λ_{k_s}, λ_v, λ_S (see (5.1))	4 (see (5.4))	0 (small scale model impossible)
	rough turbulent	4 viz: the same as above	3 (see (5.9))	1
Distorted model		5 viz: λ_x, λ_y, λ_v, λ_S, λ_c (see (5.32))	3 (see (5.32))	2

* The only flow velocity that appears in eqns. (5.26) and (5.27) is the average velocity v. Hence among all velocities of the flow it is only v which is regarded as the section property and which is thus guaranteed to be reproduced correctly in the distorted model designed according to the scale relations (5.32).

5.3.3 Scale Relations for Friction Factor and Roughness Up to this point, any operation concerning the friction factor c was carried out by using the symbolic notation

$$c = \varphi_c\left(\text{Re}; \frac{k_s}{h}\right)$$

It is now intended to express the scale of c and of the relative roughness by using the actual form of the function φ_c. From the content of 5.3.1, it is clear that the distortion should be applied only to those flows which possess sufficiently wide cross-sections to contain the central regions where the flow can be regarded as undistinguishable from two-dimensional flows. Accordingly, the following considerations will be confined to two-dimensional flows only. If the depth of a two-dimensional flow varies only gradually in the flow direction, then the logarithmic velocity distribution is valid. In this case, the average velocity v is related to the maximum velocity u_{max}, at the free surface of the two-dimensional flow by

$$\frac{v}{v_*} = \frac{u_{max}}{v_*} - 2{\cdot}5 \tag{5.34}$$

where

$$\frac{u_{max}}{v_*} = \frac{1}{\kappa} \ln \frac{h}{k_s} + B_s \tag{5.35}$$

(with $\kappa = 0{\cdot}4$ Von Karman constant)

The eqn. (5.34) follows from the universal velocity distribution law $(u_{max} - u)/v_* = -(1/\kappa) \ln y/h$. The relation (5.35) being obtained by substituting $k = k_s$, $y = h$ and $u = u_{max}$ in the logarithmic velocity distribution formula (3.28). (See the literature on fluid mechanics or boundary layer theory). In the relation (5.35) the value of B_s is the function of the Reynolds number $v_* k_s/\nu$ that is given by the experimental curve in Fig. 3.3.

Eliminating u_{max}/v_* from (5.34) and (5.35), and taking into account v/v_* implies c, we arrive at the following general expression of the friction factor

$$c = \frac{1}{\kappa} \ln \frac{h}{k_s} + (B_s - 2{\cdot}5) \tag{5.36}$$

As has been already pointed out in Section 3.3, the value of B corresponding to the transitional regime (Zone II) cannot be given by a simple analytical form. Accordingly, the following considerations, carried out using the general expression (5.36), can be regarded as those corresponding to the transitional regime (i.e. as those corresponding to $\approx 5 < v_* k_s/\nu < \approx 70$). Dividing the model and prototype versions of (5.36) we obtain

$$\lambda_c = \frac{c''}{c'} = \frac{\dfrac{1}{\kappa}\ln\dfrac{h''}{k_s''} + (B_s'' - 2{\cdot}5)}{\dfrac{1}{\kappa}\ln\dfrac{h'}{k_s'} + (B_s' - 2{\cdot}5)}$$

i.e.

$$\lambda_c = \frac{\ln\dfrac{h'}{k_s'}\cdot\dfrac{\lambda_h}{\lambda_{k_s}} + \kappa(B_s'' - 2{\cdot}5)}{\ln\dfrac{h'}{k_s'} + \kappa(B_s' - 2{\cdot}5)}$$

or

$$\lambda_c = 1 + \frac{\ln\dfrac{\lambda_h}{\lambda_{k_s}} + \kappa(B_s'' - B_s')}{\ln\dfrac{h'}{k_s'} + \kappa(B_s' - 2{\cdot}5)} \tag{5.37}$$

Since $\lambda_c^{-2} = \lambda_S = n$ (see (5.32)), eqn. (5.37) can also be given as a relation between the scales λ_h, λ_{k_s} and the distortion n.[5,6]

From (5.37) we determine the following expression for the scale of relative roughness

$$\frac{\lambda_{k_s}}{\lambda_h} = \left(\frac{k_s'}{h'}\right)^{\lambda_c - 1} . \, e^{\kappa[(B_s'' - 2\cdot 5) - \lambda_c(B_s' - 2\cdot 5)]} \tag{5.38}$$

Considering that, according to (5.32)

$$\lambda_c = \sqrt{\left(\frac{\lambda_x}{\lambda_y}\right)} = \frac{1}{\sqrt{n}}$$

we can express (5.38) as

$$\frac{\lambda_{k_s}}{\lambda_h} = \left(\frac{k_s'}{h'}\right)^{\frac{1}{\sqrt{n}} - 1} . \, e^{\kappa[(B_s'' - 2\cdot 5) - \frac{1}{\sqrt{n}}(B_s' - 2\cdot 5)]} \tag{5.39}$$

or as

$$\frac{k_s''}{h''} = \left(\frac{k_s'}{h'}\right)^{\frac{1}{\sqrt{n}}} . \, e^{\kappa[(B_s'' - 2\cdot 5) - \frac{1}{\sqrt{n}}(B_s' - 2\cdot 5)]} \tag{5.40}$$

(i) If the flow is in rough turbulent regime, that is, $v_* k_s/\nu$ is larger than

$\approx$70 (Zone III in Section 3.3), then, as follows from Fig. 3.3 (and as is indicated by the eqn. (3.31))

$$B_s = (B_s)_3 = 8{\cdot}5$$

Substituting this value of B_s in the general form (5.36), we arrive at the following expression of the friction factor corresponding to the two-dimensional rough turbulent flow with a free surface

$$c = \frac{1}{\kappa} \ln \frac{h}{k_s} + 6{\cdot}00 \tag{5.41}$$

which can also be written as

$$c = \frac{1}{\kappa} \ln \left(11 \frac{h}{k_s}\right) \tag{5.42}$$

Dividing the model and prototype versions of any of these expressions, or simply substituting

$$B_s' = B_s'' = 8{\cdot}5$$

in the general form (5.37) and taking into account that $\kappa = 0{\cdot}4$, we obtain

$$\lambda_c = 1 + \frac{\ln \dfrac{\lambda_h}{\lambda_{ks}}}{\ln \dfrac{h'}{k_s'} + 2{\cdot}4} \tag{5.43}$$

or

$$\lambda_c = 1 + \frac{\ln \dfrac{\lambda_h}{\lambda_{ks}}}{\ln \left(0{\cdot}875 \dfrac{h'}{k_s'}\right)} \tag{5.44}$$

In the case of a rough turbulent flow, the relation (5.40) becomes

$$\frac{k_s''}{h''} = \left(\frac{k_s'}{h'}\right)^{\frac{1}{\sqrt{n}}} . \, e^{2\cdot4\left(1-\frac{1}{\sqrt{n}}\right)} \tag{5.45}$$

It is remembered that the relations (5.44) and (5.45) can be used only if the model and prototype values of $v_* k_s/\nu$ (which might not necessarily be equal) are both larger than $\approx$70.

(ii) If the model and prototype are in hydraulically smooth regime, that is, if $v_* k_s/\nu$ is smaller than $\approx$5 (Zone I in the Section 3.3), then the quantity B_s is given (by (3.30)) as

$$B_s = (B_s)_1 = \frac{1}{\kappa} \ln \frac{v_* k_s}{\nu} + 5{\cdot}5 \tag{5.46}$$

(straight line S_1 in Fig. 3.3) while the friction factor c and its scale λ_c is given (independently of k_s) by

5·46 → 5·36

$$c = \frac{1}{\kappa} \ln \frac{v_* h}{\nu} + 3{\cdot}0 \tag{5.47)*}$$

and

$$\lambda_c = 1 + \frac{\ln (\lambda_h \lambda_{v_*})}{\ln \left(\dfrac{v_*' h'}{\nu'}\right) + 3{\cdot}0\kappa} \tag{5.48}$$

Considering the second condition of (5.32), and the values $\lambda_{v_*} = \lambda_y / \sqrt{\lambda_x}$, $\lambda_h = \lambda_y$ and $\kappa = 0.4$ we arrive at

$$\sqrt{\left(\frac{\lambda_x}{\lambda_y}\right)} = 1 + \frac{\ln \left(\dfrac{\lambda_y^2}{\sqrt{(\lambda_x)}}\right)}{\ln \left(\dfrac{v_*' h'}{\nu'}\right) + 1{\cdot}2} \tag{5.49}$$

It follows that in the case of a hydraulically smooth regime the scales λ_x and λ_y must satisfy a certain condition which depends on the prototype Reynolds number $v_*' h' / \nu'$ and which substitutes the second relation in (5.32). A laminar flow or a hydraulically smooth turbulent flow can hardly be present in a natural channel or river. The relations (5.48) and (5.49) have been derived here only as a formality; we will not have recourse to them again.

If the model and prototype values of $v_* k_s / \nu$ are both larger than ≈ 5, that is, if both flows are in the transitional regime or one is in the transitional regime while the other is in the rough turbulent regime, then, strictly speaking, the values B_s'' and B_s' are not equal (if the model and prototype values of $v_* k_s / \nu$ are not equal) and, as has been pointed out earlier, the scales should be determined according to the general forms (5.37) to (5.40). On the other hand, as one can see from Fig. 3.3, the quantity B_s does not vary appreciably in the semi-infinite range

$$\approx 5 < \frac{v_* k_s}{\nu} < \infty$$

the *maximum* deviation of B_s from 8.5 being

$$\frac{9{\cdot}5 - 8{\cdot}5}{8{\cdot}5} = \frac{1}{8{\cdot}5} \approx \%12$$

* The relation (5.47) is obtained by substituting (5.46) into the general form (5.36).

Hence, even if the model (and perhaps the prototype) is in the transitional regime, one can still use the simpler scale relations (5.44) and (5.45) corresponding to the rough turbulent flow (rather than the general forms (5.37) to (5.40)) at least for the first approximation. Once the required quantities are computed, one can always determine (from Fig. 3.3) the values B_s' and B_s'' corresponding to $v_*'k_s'/\nu'$ and $v_*''k_s''/\nu''$, and subsequently improve the values by repeating the calculations using the general versions (5.37) to (5.40) (successive approximation).

It should also be noted that, at present, the limiting value of k_s/h, beyond which the logarithmic velocity distribution becomes invalid, is not known to any precision. The highest value of relative roughness, k_s/h, which was used in the experimental measurements of J. Nikuradse, and for which the logarithmic velocity distribution is undoubtedly valid, is $k_s/h = 1/15$.[7] From more recent measurements by C. R. Neill (unpublished) with rough turbulent flows having a free surface, it appears that the logarithmic velocity distribution can still be regarded as valid if $k_s/h = 1/8{\cdot}35$, but not if $k_s/h = 1/5{\cdot}25$.

In the above derivations, logarithmic velocity distribution was assumed for both model and prototype. Although we are here concerned with models where only the section properties are supposed to be adequately reproduced, it would nevertheless be desirable if model and prototype velocities along y varied according to the same law. For this reason, when model scales are being selected, care should be taken so that the value of relative roughness in the model should be less than, say, 1/5 or 1/7. To be on the safe side, 1/10 should be adopted as the maximum value of model relative roughness (when dealing with gradually varying flows).

The foregoing principles will now be demonstrated in some numerical examples.

EXAMPLE 5.1

Let us first consider as a prototype that lowland river which was used in Subsection 5.2.2 as an example. The characteristics of the prototype are

$$h' = 5{\cdot}00 \text{ m};\ S' = 0{\cdot}0001;\ k_s' = 0{\cdot}02 \text{ m}$$
$$(g = 9{\cdot}81 \text{ m/s}^2;\ \nu = 10^{-6} \text{ m}^2/\text{s})$$

In addition to these characteristics, we introduce the river width

$$B' = 200 \text{ m}$$

We will attempt to design a distorted model according to the method described in 5.3.1, and thus according to the scale relations (5.32). *Any* two scales of these scale relations can be selected freely, and therefore we can start, for example, by selecting λ_y and λ_{k_s}. Accordingly, we chose freely the model flow depth h'' and the size k_s'' of the model roughness, say, as

$$h'' = 167 \text{ mm} \quad \text{and} \quad k_s'' = 10 \text{ mm}$$

(Since $k_s''/h'' = 1/16{\cdot}7$, the condition $k_s/h < 1/10$ is satisfied.) Hence, the selected two scales are

$$\lambda_y = \frac{167}{5000} = \frac{1}{30} \quad \text{and} \quad \lambda_{k_s} = \frac{1}{2}$$

All the remaining scales are now determined, and can be computed from the related equations. In 5.2.2 we have seen that our prototype corresponds to $v_*'k_s'/\nu = 1400 > \approx 70$, and thus that the prototype flow is rough turbulent. Assuming that the model flow is also such, we compute the model and prototype values of the friction factor according to (5.41) as follows

$$c'' = 2{\cdot}50 \ln \left(\frac{167}{10}\right) + 6{\cdot}00 = 13{\cdot}54$$

$$c' = 2{\cdot}50 \ln \left(\frac{5000}{20}\right) + 6{\cdot}00 = 20{\cdot}30$$

Thus

$$\lambda_c = \frac{13{\cdot}54}{20{\cdot}30} = \frac{1}{1{\cdot}5}$$

Using this value in the second eqn. of (5.32) we obtain

$$\frac{\lambda_y}{\lambda_x} = n = 1{\cdot}5^2 = 2{\cdot}25$$

and consequently

$$\lambda_x = \frac{1}{2{\cdot}25 \times 30} = \frac{1}{67{\cdot}5}$$

Hence, the river width in the model is

$$B'' = \frac{200}{67{\cdot}5} = 2{\cdot}97 \approx 3 \text{ m}$$

while the model (width/depth) ratio is

$$\frac{B''}{h''} = \frac{300}{16{\cdot}7} = 18$$

The central region of the model extends to

$$B_c'' \approx B'' - 5h'' = 2{\cdot}97 - 5 \times 16{\cdot}7 = 2{\cdot}035 \approx 2 \text{ m}$$

and thus it forms two-thirds of the model width, which is perfectly reasonable. From the third equation of (5.32) we obtain for the velocity scale

$$\lambda_v = \sqrt{(\lambda_y)} = \frac{1}{\sqrt{30}} = \frac{1}{5{\cdot}48}$$

In the case of a distorted model it is not possible to give a single formula (such as (5.10)) in order to indicate how the scale of any property A of the flow has to be determined. Indeed, the composition of two geometric scales λ_x and λ_y in the expression of the scale of a quantity depends entirely on the physical meaning of that quantity. For example, the flow rate Q' is the product of the cross-sectional area with the average velocity v, i.e.

$$\lambda_Q = \lambda_v \lambda_\omega$$

Here, the area ω is perpendicular to the direction x of the flow, and thus its measure involves both x and y. Accordingly, we have

$$\lambda_\omega = \lambda_x \lambda_y$$

(as implied by the second equation in (5.30)), and consequently

$$\lambda_Q = \lambda_v \lambda_x \lambda_y = \sqrt{(\lambda_y)}\lambda_x \lambda_y = \lambda_x \lambda_y^{\frac{3}{2}}$$

Consider now the shear force T acting on an area $\bar{\omega}$ of the surface of the bed. The scale of this force is given by the product

$$\lambda_{\mathrm{T}} = \lambda_{\tau_0} \lambda_{\bar{\omega}}$$

where the scale of $\tau_0 = \gamma S h$ is given by

$$\lambda_{\tau_0} = \frac{\lambda_y^2}{\lambda_x}$$

However, in this case the co-ordinate y has no bearing on the measure of the area $\bar{\omega}$, and therefore

$$\lambda_{\bar{\omega}} = \lambda_x^2$$

Using the values of λ_{τ_0} and $\lambda_{\bar{\omega}}$ in the expression of λ_{T} we arrive at

$$\lambda_{\mathrm{T}} = \lambda_y^2 \lambda_x$$

Observe that λ_x, λ_v and λ_ρ are the scales of three dimensionally independent parameters, and thus the scale λ_A of any property A can always be regarded as given by a power product of λ_x, λ_v and λ_ρ. On the other hand, since $\lambda_\rho = 1$, while $\lambda_v = \sqrt{(\lambda_y)}$, a power product of λ_x, λ_v and λ_ρ is virtually a power product of only λ_x and λ_y. This is the reason for λ_Q and λ_T being given above as the functions of λ_x and λ_y alone.

It remains to check whether the model flow is, indeed, in rough turbulent regime as was assumed in the preceding calculations. According to the first eqn. of (5.32) we have

$$S'' = \frac{\lambda_y}{\lambda_x} \cdot S' = 2{\cdot}25 \times 0{\cdot}0001 = 0{\cdot}000225$$

$$v_*'' = \sqrt{(gh''S'')} = \sqrt{(9{\cdot}81 \times 16{\cdot}7 \times 0{\cdot}000225)}$$
$$= 0{\cdot}0192 \text{ m/s} \approx 0{\cdot}02 \text{ m/s}$$

and thus

$$\frac{v_*''k_s''}{\nu} \approx \frac{0{\cdot}02 \times 0{\cdot}01}{10^{-6}} = 200 > \approx 70$$

i.e. the assumption made is valid.

Hence, the small scale model which could not have been designed according to scale relations (5.9) can be designed, without difficulty, by the method implied by the relationships (5.32).

EXAMPLE 5.2

In practice, one does not usually determine the friction factor c by means of equations such as (5.36), for the value of k_s is often unknown. Indeed,

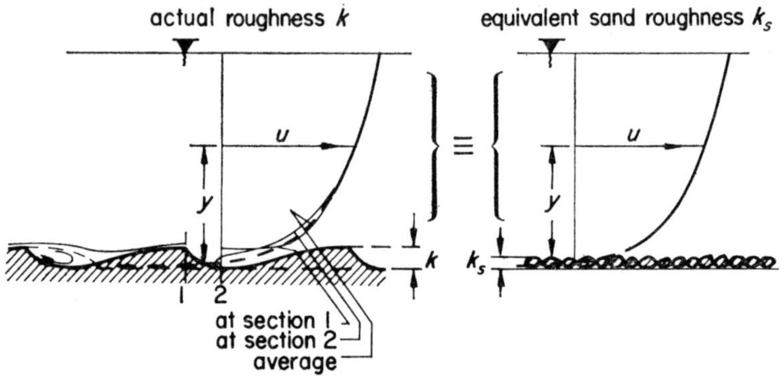

FIG. 5.3

it is only if the river or channel possesses a rough plain bed consisting of tightly packed, equal size, grains that one can, with reasonable confidence, assert that the value of k_s is known, and that it is approximately equal to the grain size. In many cases, however, the geometry of the bed's irregularities does not resemble the geometry of the granular sand roughness at all (Fig. 5.3). In such cases, k_s cannot be determined by a direct method, and thus the friction factor c cannot be computed from eqn. (5.36). Indeed, one frequently attempts to determine the value of c and then, if necessary, to compute k_s from eqn. (5.36).

According to definition, the prototype value of the friction factor is given by

$$c' = \frac{v'}{v_*'}$$

and thus

$$c' = \frac{Q'/\omega'}{\sqrt{(R'J'g)}}$$

Knowing from field measurements the flow rate Q', the value of the energy gradient J', and the geometric properties ω' and R' of a typical cross-section, one can determine the value c' from the relation above. However,

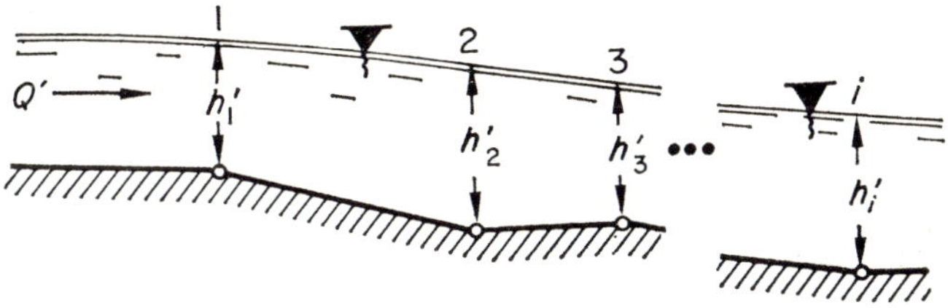

FIG. 5.4

even this approach can hardly be regarded as reliable. Indeed, it is difficult to estimate which cross-section of the gradually varying flow should be regarded as typical, or which value of the gradually varying gradient J' should be regarded as representative.

Therefore, in practice, one often prefers to avoid the risk of determining c' and/or k_s' altogether. Instead, one determines the flow rate Q' and observes the position of free surface corresponding to that flow rate (for the region of the river that will be subjected to development). If the position of the free surface, or to be more precise if the longitudinal section of the river, is provided, then a series of prototype flow depths h_i', $(i = 1, 2, \ldots)$ corresponding to a flow rate Q' is known (Fig. 5.4). Usually one can have the free surface profiles corresponding to a number of flow rates Q_j $(j = 1, 2, \ldots m)$. The knowledge of Q_j' and h_{ij}' in addition to the information on the topography (geometry) of the river bed is usually sufficient to design and build a model (without knowing the values k_s' and c'). The purpose of the present example is to illustrate how this is done.

Let us first assume that a single flow rate is given, which is

$$Q' = 2010 \text{ m}^3/\text{s}$$

while the corresponding flow depths h' in, say, five sections i are

i	1	2	3	4	5
h_i' (m)	6·80	6·92	7·03	7·20	7·36

We begin with the selection of two scales. For example, we chose two geometric scales λ_x and λ_y, as follows

$$\lambda_x = \frac{1}{180} \quad \text{and} \quad \lambda_y = \frac{1}{60}$$

The rest of the scales must follow from the related equations. Using these values of λ_x and λ_y, we determine the following value for the flow rate scale

$$\lambda_Q = \lambda_x \ \lambda_y^{\frac{3}{2}} = \frac{1}{180 \times 60^{3/2}} = \frac{1}{83\,600}$$

Thus, thc model flow rate is

$$Q'' = \frac{2010}{83\,600} = 0{\cdot}024 \text{ m}^3/\text{s} = 24 \text{ litres/s}$$

The roughness of the model flow boundaries must be such that the model flow rate $Q'' = 24$ litres/s should produce the following model flow depths h_i'' (determined as $h_i'/60$) in the corresponding sections

i	1	2	3	4	5
h_i'' (cm)	11·34	11·52	11·72	12·00	12·29

The required roughness is achieved by trial and error experiments (often referred to as the 'calibration of model'). The model roughness elements, concrete cubes, little stones, etc. are fixed (glued) on the rigid concrete floor of the model according to a certain configuration (Plate 2a). Subsequently, this first trial configuration undergoes adjustment until the model flow rate Q'' yields, in the sections i, the required flow depths h_i''. Sometimes, instead of the roughness elements mentioned, above $\approx$6 mm diameter metal rods are used. These rods are fixed on the model bed vertically, and their length is selected so as to extend up to the free surface. In this case the adjustment consists of determining the appropriate number of rods placed per square foot of the model bed. The advantage of this method is that it acts on the whole body of the flow rather than on its perimeter only; hence a more natural (more prototype-like) distribution of model velocities can be achieved.

The model roughness can be adjusted more realistically if more than one prototype flow rate Q_j' and the corresponding series of the prototype flow depths h_{ij}' are given. Let $Q'_{\min}$ and $'_{\max}$ be the smallest and largest among the given flow rates Q_j'. It can happen that a certain uniform distribution of the model roughness might yield the required model flow depths, e.g. for $Q''_{\min}$ (and for other small flow rates Q_j''), but not for $Q''_{\max}$ (nor for other large Q_j''), or vice versa. This is an indication that the prototype roughness is not uniform; the value of the bed roughness is different from that of the bank roughness. In such cases, the model roughness must be formed accordingly. First, the bed roughness should be adjusted for the smallest flow rate $Q''_{\min}$; subsequently, one should try to achieve the agreement with the intermediate flow rates and $Q''_{\max}$ by adjusting mainly the bank roughness.

In the method described above, neither the scale of roughness nor that of the friction factor was used. Yet the model designed according to the method described is as adequate with respect to the frictional losses as that implied by the set (5.32). To demonstrate the validity of this statement, it is sufficient to show that the expression

$$\lambda_c = \frac{\lambda_Q}{\lambda_\omega \lambda_R^{\frac{1}{2}} \lambda_J^{\frac{1}{2}}}$$

which follows from

$$c = \frac{Q/\omega}{\sqrt{(RJg)}}$$

is equivalent to the second condition in (5.32). The distorted model described in the present example obviously satisfies the conditions

$$\lambda_\omega = \lambda_x \lambda_y,\ \lambda_Q = \lambda_x \lambda_y^{\frac{3}{2}},\ \lambda_R \approx \lambda_y,\ \lambda_J = \lambda_y \lambda_x^{-1}$$

Substituting these values in the expression of λ_c above, we arrive at

$$\lambda_c = \sqrt{\left(\frac{\lambda_x}{\lambda_y}\right)}$$

which is the second condition of (5.32).

EXAMPLE 5.3

From eqn. (5.40) or (5.45), it is clear that, under the equality of the remaining characteristics, the model relative roughness k_s''/h'' is an increasing function of the prototype relative roughness k_s'/h'. It is intended to reveal what can be the largest possible value of the prototype relative roughness k_s'/h' that can be reproduced in a model which possesses the (highest permissible) roughness

$$\frac{k_s''}{h''} = \frac{1}{10}$$

and which has the distortion

$$n = 1;\ 2;\ 3;\ 4;\ 5$$

From the content of the problem, it is clear that the model and prototype flows should be treated as rough turbulent flows. Accordingly, substituting $k_s''/h'' = 1/10$ in (5.45), we arrive at the relation between k_s'/h' and n

$$\frac{1}{10} = \left(\frac{k_s'}{h'}\right)^{\frac{1}{\sqrt{n}}} \times e^{2\cdot4\left(1-\frac{1}{\sqrt{n}}\right)}$$

From this relation, we determine the following values of k_s'/h' corresponding to the given distortions n

n	1	2	3	4	5
$\frac{k_s'}{h'}$	$\frac{1}{10}$	$\frac{1}{70}$	$\frac{1}{310}$	$\frac{1}{1100}$	$\frac{1}{3300}$

Consider, for example, the value $k_s'/h' = 1/3300$ corresponding to $n = 5$. This value implies that if the flow depth is $h' = 10$ m, then the size k is only 3 mm. The prototype corresponding to such specifications can hardly be regarded as 'common' in engineering practice. Indeed, such an unusual prototype can be, for example, a very large concrete channel (forebay of a hydroelectric power plant) or a large river having such a tranquil flow (small slope) that it is not capable of transporting sediment larger than 3 mm grain size and so can be assumed to possess a plain granular bed of $k_s' = 3$ mm. If sediment transport were taking place then the bed, in general, would be covered by sand waves, which under the given conditions would correspond to a value of k_s much larger than 3 mm. The range of slopes S of the river which possesses the required characteristics, and yet which is unable to transport the sediment of 3 mm diameter can be determined as follows. Since the flow is rough turbulent

$$\frac{v_* k_s}{\nu} \approx \frac{\sqrt{(gSh)}k_s}{\nu} = \frac{\sqrt{(9\cdot81 \times S \times 10)}0\cdot003}{10^{-6}} > \approx 70$$

and thus

$$S > 5{\cdot}45 \times 10^{-6}$$

is valid. On the other hand, since the sediment is not transported (by the rough turbulent flow) we have, according to the Shields curve (introduced in the next chapter)

$$\frac{\rho v_*^2}{\gamma_s k_s} \approx \frac{\gamma}{\gamma_s} \cdot \frac{hS}{k_s} = \frac{1}{1{\cdot}65} \frac{10 \times S}{0{\cdot}003} < 0{\cdot}05 \text{ and thus } S < 2{\cdot}48 \times 10^{-5}$$

where γ_s is the specific weight of the gravel in water. Hence the range S is

$$5{\cdot}45 \times 10^{-6} < S < 2{\cdot}48 \times 10^{-5}$$

Here it was assumed that the flow is uniform and thus that the energy gradient $J = -\partial H/\partial x$ coincides with the slope S of the channel (see the text following eqn. (5.19)).

On the other hand, one can frequently encounter river models having a distortion of 5 or even 10 or more. Furthermore, some of these models agree very well with their prototypes which do not necessarily possess particularly small roughnesses k_s'/h'. The reason for such agreement (which appears to be in disagreement with the numerical values determined in the present example) lies in the fact that the flows in these models (and prototypes) are not gradually varying flows, and thus they do not belong to the category of the flows considered in the present subsection. This brings us to the subject to be considered next.

5.3.4 Non-uniform River and Channel Flows (not necessarily 'gradually varying') In this chapter, it has been assumed that the non-uniform flow under consideration is a gradually varying (along x) flow. This implies that even though in the equation of motion (5.26), the derivatives of some quantities with respect to x (namely $\partial\omega/\partial x$, dh/dx) are present, the values of these derivatives have to be regarded as small. If the flow depth h, the cross-section of the flow, and the slope S of the channel bed vary only gradually while the curvature of the channel axis is negligible, then it is, of course, only natural to assume that the energy loss (per unit length of the flow) is entirely due to friction, that is, that the following relation

$$-\frac{dH}{dx} = \frac{1}{c^2}\frac{v^2}{gR} = \frac{1}{c^2}\,\mathrm{Fr} \qquad (5.50)$$

is valid (as was postulated during the derivation of (5.26)).* However, if the flow cannot be regarded as gradually varying, that is, if even one of

* See the derivation of (5.26), i.e. of (1.92) in Example 1.4.

the geometric characteristics mentioned varies appreciably, then it becomes impossible to claim that the energy loss between two sections of the flow (and consequently the energy loss related to the unit length of the channel) will be entirely due to friction. Indeed, as is well-known, sudden enlargement or sudden contraction of the flow cross-section, an appreciable variation in the depth or width along a relatively short distance, even if the cross-sectional area remains the same, the presence of a bend, a discontinuous change in the slope of the channel bed, and so on, all give rise to certain energy losses. In the following text, these energy losses due to 'non-gradual' changes in the flow geometry, will be referred to as local energy losses.* It is also well-known that the local energy losses (per unit weight of the fluid) are proportional to the velocity head $v^2/2g$. Let

$$\zeta_i \frac{v^2}{2g} \tag{5.51}$$

be the energy loss corresponding to the particular geometric factor i. In this case, the cumulative energy loss due to N geometric factors that are present in the river region of the length L, can be expressed as follows

$$\left(\sum_{i=1}^{N} \zeta_i\right) \frac{v^2}{2g} \tag{5.52}$$

where v is a typical velocity of the region L. Dividing (5.52) by L, we arrive at the following expression of the (average) energy loss per unit length (of the region L) caused by the changes in the flow geometry

$$\frac{1}{2}\left(\sum_{i=1}^{N} \zeta_i\right) \frac{v^2}{gL} = \zeta_L \frac{v^2}{gL} \tag{5.53}$$

$$\left(\text{with } \zeta_L = \frac{1}{2} \sum_{i=1}^{N} \zeta_i\right)$$

Hence, if the flow is not necessarily gradually convergent, then the value of $\mathrm{d}H/\mathrm{d}x$ is given by the following sum of the frictional losses and the losses due to the changes in the geometry (local energy losses)

$$-\frac{\mathrm{d}H}{\mathrm{d}x} = \zeta_L \frac{v^2}{gL} + \frac{1}{c^2} \frac{v^2}{gR} \tag{5.54}$$

* Here we have deliberately departed from a rather unfortunate English term 'minor losses'. In a system consisting of long straight pipes and few bends, the losses mentioned are indeed minor losses (in comparison to the major friction losses). However, in a system with perpetually and intensively varying geometry, the minor losses become, in fact, major losses while friction losses turn into minor losses. The term local energy losses adopted in the present text is the literal translation of the Russian and German terms 'местные потери энергии' and 'örtliche Energieverluste', which are indentical in meaning.

* $\frac{h}{l} = \frac{f v^2}{8gR} = \frac{1}{c^2}\frac{v^2}{gR}$; $\therefore \frac{f}{8} = \frac{1}{c^2}$, $c = \sqrt{\frac{8}{f}}$

which can be written as

$$-\frac{\mathrm{d}H}{\mathrm{d}x} = \left(\zeta_L \frac{R}{L} + \frac{1}{c^2}\right) \mathrm{Fr} = E\,\mathrm{Fr} \tag{5.55}$$

The relation (5.55) indicates clearly that the energy loss $\mathrm{d}H/\mathrm{d}x$ per unit length of the river region L will be mainly due to friction, i.e. that (5.55) will be reduced into (5.50) if the variation in the geometry along the region L is only feeble (gradually varying flow). Conversely, the loss will be mainly due to the local energy losses if the flow geometry varies with a sufficient intensity and/or frequency (i.e. if ζ_i and/or N which determine the value of ζ_L is/are sufficiently large). Using (5.55)) (instead of (5.50)) and following the same pattern of derivation (shown in Example 1.4), we arrive at

$$\frac{1}{S'}\frac{\mathrm{d}h'}{\mathrm{d}x'} = \frac{1 - \dfrac{\mathrm{Fr}'}{S'}\left[\left(\dfrac{1}{c'^2} + \zeta_L{}'\dfrac{R'}{L'}\right) - \dfrac{\alpha}{\chi'}\dfrac{\partial\omega'}{\partial x'}\right]}{1 - \alpha\dfrac{B'}{\chi'}\mathrm{Fr}'} \tag{5.56}$$

The form (5.56) is more general than its classical counterpart (5.26) (or (1.92)) in the sense that the non-uniform flow described by (5.56) need not necessarily be gradually varying (the round bracket in the numerator being the multiplier E in the expression (5.55)). Since the relation (5.56) differs from (5.26) (or (1.92)) only because of the term

$$\zeta_L{}'\frac{R'}{L'}$$

the scale relations which must follow from the model and prototype identity of (5.56) can differ from those of (5.32) only because of the additional condition imposed by the term above.

Let us determine this additional condition. Since the scale of each term in the square brackets in (5.56) must obviously be equal to the same ratio λ_y/λ_x (implying the distortion n), we have

$$\lambda_{\zeta_L}\frac{\lambda_R}{\lambda_L} = \frac{\lambda_y}{\lambda_x} \tag{5.57}$$

On the other hand, considering that

$$\lambda_R = \frac{\lambda_\omega}{\lambda_\chi} \approx \frac{\lambda_x\lambda_y}{\lambda_x} = \lambda_y$$

and that

$$\lambda_L = \lambda_x$$

we arrive at

$$\lambda_{\zeta_L} \approx 1 \tag{5.58}$$

Hence, the fulfilment of (5.57) is equivalent to the fulfilment of (5.58).

The coefficients ζ_i of the local energy losses are usually the functions of certain dimensionless ratios describing the geometry responsible for the local energy losses. It appears that the mentioned ratios are often identical in model and prototype, in spite of the distortion, and thus, that the coefficients ζ_i can also be regarded as identical. Consequently, the condition (5.57) can frequently be regarded as satisfied automatically. In order to demonstrate this, let us consider, for example, the energy losses due to

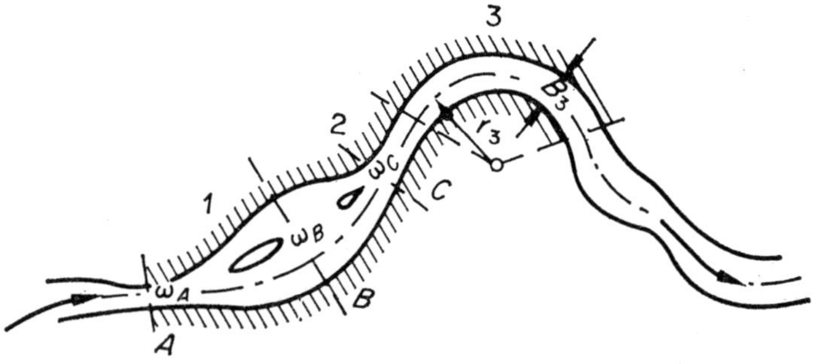

FIG. 5.5

sudden expansion, sudden contraction, and the presence of a bend. These three particular losses usually constitute the bulk of all local energy losses. The coefficients ζ_1, ζ_2 and ζ_3 of the three mentioned losses can always be considered as the following functions of the 'local geometry'

$$\left.\begin{aligned}\zeta_1 &= \varphi_1(X_1)\\ \zeta_2 &= \varphi_2(X_2)\\ \zeta_3 &= \varphi_3(X_3)\end{aligned}\right\} \tag{5.59}$$

where X_1, X_2 and X_3 stand for the ratios

$$X_1 = \frac{\omega_A}{\omega_B} \quad X_2 = \frac{\omega_B}{\omega_c} \quad X_3 = \frac{B_3}{r_3} \tag{5.60}$$

The meaning of the characteristics on the right hand sides being clear from the sketch in Fig. 5.5. For example, since

$$\omega_A'' = \lambda_x \lambda_y \omega_A' \quad \text{and} \quad \omega_B'' = \lambda_x \lambda_y \omega_B'$$

it is clear that the model and prototype values of X_1 are identical

$$X_1'' = \frac{\omega_A''}{\omega_B''} = \frac{\omega_A'}{\omega_B'} = X_1'$$

In a similar manner, one can show that the model and prototype values of X_2 and X_3 are also identical. But if the values of the variables of the functions ζ_1, ζ_2 and ζ_3 are identical, then the values of these functions themselves, and thus of their sum

$$\zeta_L = \zeta_1 + \zeta_2 + \zeta_3$$

can also be regarded as identical in both model and prototype. Hence, the condition (5.58), and thus (5.57), can be regarded as satisfied (at least with respect to the most important local losses implied by ζ_1, ζ_2 and ζ_3). We say *can* also be regarded as identical rather than *must* also be identical, because we cannot exclude the possibility that the form of any of the functions in (5.59) might, to a certain degree (however small), be affected as a result of the change in the shape of the flow cross-sections (due to the distortion).

From these considerations, it is clear that if the non-uniform flow is not subjected to the restriction of being gradually varying, then its similarity should be considered according to the following generalized version of (5.32)

$$\lambda_S = \frac{\lambda_y}{\lambda_x} \qquad \lambda_E = \frac{\lambda_y}{\lambda_x} \qquad \lambda_{\mathrm{Fr}} = 1 \tag{5.61}$$

The set (5.61) differs from (5.32) because the consideration of $1/\lambda_c^2 = \lambda_y/\lambda_x$ is substituted by that of a more general relation

$$\lambda_E = \frac{\zeta_L'' \dfrac{R''}{L''} + \dfrac{1}{c''^2}}{\zeta_L' \dfrac{R'}{L'} + \dfrac{1}{c'^2}} = \frac{\lambda_y}{\lambda_x} \tag{5.62}$$

i.e.

$$\lambda_E = \frac{\zeta_L' \dfrac{R'}{L'} \left[\lambda_{\zeta_L} \dfrac{\lambda_R}{\lambda_L}\right] + \dfrac{1}{c'^2} \left[\dfrac{1}{\lambda_c^2}\right]}{\zeta_L' \dfrac{R'}{L'} + \dfrac{1}{c'^2}} = \frac{\lambda_y}{\lambda_x} \tag{5.63}$$

Assuming, in accordance with the explanations above, that the conditions (5.58) and (5.57) are satisfied, and thus that $\lambda_{\zeta_L} = 1$, and taking into account that λ_R/λ_L is equal to λ_y/λ_x, the second condition of the set (5.61) can be expressed in the following (open) form

$$\frac{\zeta_L' \dfrac{R'}{L'} + \dfrac{1}{c'^2}\left[\dfrac{\lambda_x}{\lambda_y \lambda_c^2}\right]}{\zeta_L' \dfrac{R'}{L'} + \dfrac{1}{c'^2}} = 1 \tag{5.64}$$

Consider the following extreme cases:

(i) *Gradually varying flow*
In this case, the local energy losses in the river region L' are negligible in comparison to the friction losses in the same region, which implies that $\zeta_L' R'/L'$ is negligible with respect to $1/c'^2$, and thus that (5.64) reduces into

$$\frac{\lambda_x}{\lambda_y \lambda_c^2} = 1 \quad \text{or} \quad \lambda_c^2 = \frac{\lambda_x}{\lambda_y} \tag{5.65}$$

Hence, the set (5.61) satisfies the requirement of being indistinguishable from (5.32) if the non-uniform flow varies gradually.

(ii) *Very intensively varying non-uniform flows*
In this case, the friction losses along the river region L' can be neglected in comparison to the local losses in the same region, which implies that the term $1/c'^2$ is negligible compared to $\zeta_L' R'/L'$ and that the condition (5.64) reduces into the identity

$$\zeta_L' \frac{R'}{L'} \equiv \zeta_L' \frac{R'}{L'}$$

Hence, the more intensive the variation of the non-uniform flow (along x), the more the set (5.61) approaches its following limit, which no longer depends on the condition imposed by the friction factor c

$$\lambda_S = \frac{\lambda_y}{\lambda_x}, \quad \lambda_{\mathrm{Fr}} = 1 \tag{5.66}$$

Hence, the sets (5.32) and (5.66) are two opposite extremes (two special cases) of the set (5.61) representing the general case. As far as the geometric interpretation is concerned, the flow boundary (river bed) is a *surface* which separates the flow from the ground. From the considerations given here, it follows that the more irregular the shape of the mentioned surface, the less important the influence of its roughness (or of the friction factor c which is a function of the dimensionless roughness). This explains why some irregular rivers can still be reasonably well reproduced in models designed without observing the restrictions imposed by the criteria of roughness or *friction factor* (see text at end of Example 5.3).

At this stage one should perhaps pause and consider the following question, which involves what might be called the philosophy of hydraulics: Where is the upper limit of 'roughness', or, which is the same thing, the lower limit of what is referred to as 'irregularity of the flow boundaries'? Clearly, there is no such natural boundary where one could erect a post with the signs ←'roughness, 'boundary→ irregularities'. Indeed, one arrives from what can certainly be regarded as boundary irregularities at what can equally certainly be regarded as roughness by a continuous (topological) deformation process (Fig. 5.6). Accordingly, the word roughness is in a

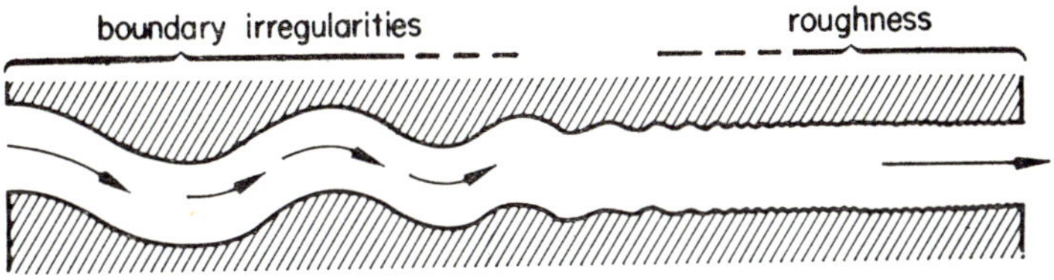

FIG. 5.6

sense a synonym for small boundary irregularities, while the term boundary irregularities is merely a different way of saying large roughness. But if so, that is, if roughness and boundary irregularities are essentially the same, why are the mathematical expressions of the energy losses due to them as different as

$$\left(\frac{1}{2}\sum_{i=1}^{N}\zeta_i\right)\frac{v^2}{gL} \quad \text{and} \quad \frac{1}{c^2}\frac{v^2}{gR}$$

The answer to this question lies in the fact that the expressions above are only *apparently* different; essentially they are the same. Indeed, since c is a certain function of k/h alone, the term implying the frictional loss (per unit length of the flow) can be expressed as follows

$$\bar{\varphi}_c\left(\frac{k}{h}\right)\frac{v^2}{gR}\text{*} \tag{5.67}$$

Let us now consider the flow in a conduit possessing an irregular flow boundary such as, for example, that shown in Fig. 5.7. We assume that

* For the sake of simplicity, the demonstration is restricted to rough turbulent flows only. Hence, the function $\bar{\varphi}_c$ is independent of $v_* k/\nu$.

this conduit is circular, and that the energy loss is due to the local expansions and contractions only. Thus, the local energy loss (along L) is given by

$$(\zeta_1 + \zeta_2)\frac{v^2}{2g}$$

and per unit length by

$$\tfrac{1}{2}(\zeta_1 + \zeta_2)\frac{v^2}{gL} \tag{5.68}$$

Considering that

$$\zeta_1 = \varphi_1\left(\frac{\omega_A}{\omega_B}\right) \quad \text{while} \quad \zeta_2 = \varphi_2\left(\frac{\omega_B}{\omega_A}\right)$$

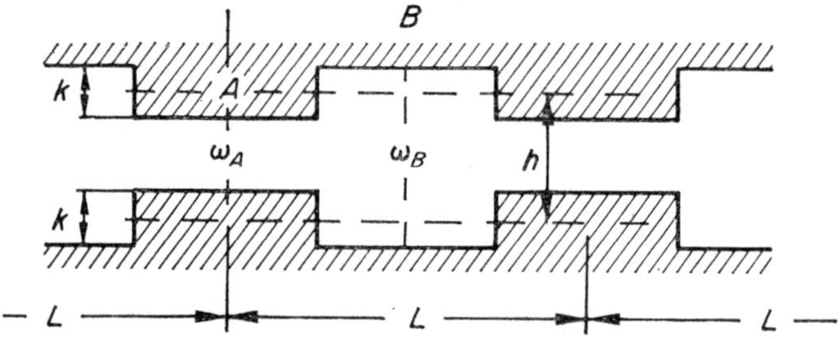

FIG. 5.7

we obtain

$$\tfrac{1}{2}(\zeta_1 + \zeta_2) = \tfrac{1}{2}\varphi_1\left(\frac{\omega_A}{\omega_B}\right) + \tfrac{1}{2}\varphi_2\left(\frac{\omega_A}{\omega_B}\right) = \varphi\left(\frac{\omega_A}{\omega_B}\right) \tag{5.69}$$

and thus

$$\varphi\left(\frac{\omega_A}{\omega_B}\right)\frac{v^2}{gL} \tag{5.70}$$

as (5.68). But

$$\omega_A \sim (h - \alpha k)^2 \quad \text{while} \quad \omega_B \sim (h + (1 - \alpha)k)^2 \tag{5.71}$$

(where α is a constant smaller than unity), and therefore

$$\frac{\omega_A}{\omega_B} = \left(\frac{h - \alpha k}{h + (1 - \alpha)k}\right) = \left(\frac{1 - \alpha\dfrac{k}{h}}{1 + (1 - \alpha)\dfrac{k}{h}}\right) = \bar{\psi}\left(\frac{k}{h}\right) \tag{5.72}$$

On the other hand, since the geometry of the boundary irregularities is specified, the length L is related to any of the linear characteristics of the flow boundary, and thus to the hydraulic radius R, by a constant proportionality. Thus

$$L = \text{const} \,.\, R \tag{5.73}$$

Substituting (5.72) and (5.73) in (5.70) we arrive at

$$\psi\left(\frac{k}{h}\right)\frac{v^2}{gR} \tag{5.74}$$

$$\left(\text{with } \psi\left(\frac{k}{h}\right) = \varphi\left[\bar{\psi}\left(\frac{k}{h}\right)\right]\Big/\text{const}\right)$$

which, in principle, is the same as (5.67).

It seems that the apparent difference in the expressions used for frictional and local energy losses is simply due to the fact that these expressions have been developed independently by different authors using their own different approaches and notations. These authors, in their own times, perhaps were not even aware that they were contributing to essentially the same subject, and thus they cannot be held responsible for the resulting heterogeneity in mathematical expressions. This, however, is a poor excuse for the fact that these (essentially the same) expressions are still persistently used in the literature in a non-unified manner, which certainly does not help us to see hydraulics as a homogeneous logical whole.

5.4 Non-stationary Flows

If the flow is not in a steady state, then the time t is an additional parameter while the Strouhal number

$$\text{St} = \frac{Ut}{l}$$

is an additional variable. Hence, the additional scale relation

$$\frac{\lambda_U \lambda_t}{\lambda_l} = 1 \tag{5.75}$$

must be satisfied.

(i) *Undistorted model*
In this case, we have only one model scale λ_l; the velocity scale being

$$\lambda_U = \sqrt{(\lambda_l)}$$

Accordingly, from the Strouhal criterion (5.75) we obtain the following value for the time scale

$$\lambda_t = \frac{\lambda_l}{\lambda_U} = \sqrt{(\lambda_l)} \tag{5.76}$$

Any variation in the prototype flow which takes place during the time t' must take place in the model during the time $t'' = t'\sqrt{(\lambda_l)}$. From the relation (5.76) it follows that in an undistorted Froudian model the time scale is equal to the velocity scale.

(ii) *Distorted model*

In this case, we have two model scales λ_x and λ_y, and thus the situation is more complicated. Consider the prototype and model flows schematically

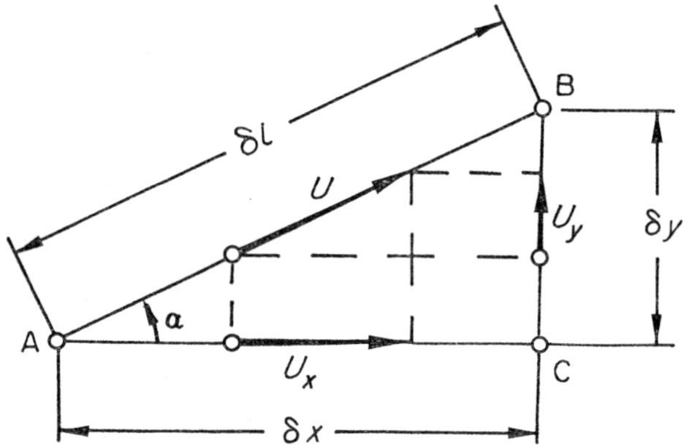

FIG. 5.8

shown in Fig. 5.8. Let us assume that during the time interval $\delta t'$ the prototype fluid particle is displaced from the point A′ to the point B′, while the model particle travels from the corresponding point A″ to the corresponding point B″ in the time $\delta t''$. Denoting the velocities:

$$\begin{aligned} &\text{along } \overline{AB} = \delta l \text{ by } U \\ &\text{along } \overline{AC} = \delta x \text{ by } U_x, \text{ and} \\ &\text{along } \overline{CB} = \delta y \text{ by } U_y \end{aligned}$$

and considering that the resultant motion along δl can always be considered by means of the components of this motion along δx and δy, we can write for prototype and model

$$\left.\begin{aligned}\delta t' &= \frac{\delta l'}{U'} = \frac{\delta x'}{U_x'} = \frac{\delta y'}{U_y'}\\ &\text{and}\\ \delta t'' &= \frac{\delta l''}{U''} = \frac{\delta x''}{U_x''} = \frac{\delta y''}{U_y''}\end{aligned}\right\} \tag{5.77}$$

respectively. From the equalities above, it follows that

$$\frac{\delta t''}{\delta t'} = \lambda_t = \text{const}$$

and consequently

$$\frac{\lambda_l}{\lambda_U} = \frac{\lambda_x}{\lambda_{U_x}} = \frac{\lambda_y}{\lambda_{U_y}} = \lambda_t = \text{const} \tag{5.78}$$

must be valid. The last equality can be brought into the following Strouhal number form

$$\lambda_{\text{St}} = \frac{\lambda_U \lambda_t}{\lambda_l} = \frac{\lambda_{U_x}\lambda_t}{\lambda_x} = \frac{\lambda_{U_y}\lambda_t}{\lambda_y} = 1 \tag{5.79}$$

The Froude number is formed by a typical velocity (say average velocity v) that is in the direction x of the flow (the typical length (R) which appears in the Froude number being perpendicular to the flow). Hence, with respect to the distorted model, the Froudian velocity scale must be interpreted as

$$\lambda_v = \lambda_{U_x} = \sqrt{(\lambda_y)} \tag{5.80}$$

Substituting this value in the relation (5.78) or (5.79) we obtain the following value for the time scale of a distorted model

$$\lambda_t = \frac{\lambda_x}{\sqrt{(\lambda_y)}} \tag{5.81}$$

which can also be expressed as

$$\lambda_t = \frac{1}{n}\sqrt{(\lambda_y)} \tag{5.82}$$

where n is the distortion. Eliminating λ_t between (5.82) and (5.79) we obtain the velocity scales

$$\lambda_{U_y} = n\sqrt{(\lambda_y)} \tag{5.83}$$

and

$$\lambda_U = n\frac{\lambda_l}{\sqrt{(\lambda_y)}} \tag{5.84}$$

In the latter expression, the scale λ_l depends on the direction l, that is, on the angle α of inclination (see Fig. 5.8). Indeed, considering that

$$\delta l = \frac{\delta x}{\cos \alpha} = \delta x \sqrt{(1 + \tan^2 \alpha)}$$

where

$$\tan \alpha = \frac{\delta y}{\delta x}$$

we determine

$$\lambda_l = \lambda_x \sqrt{\left(\frac{1 + n^2 \tan^2 \alpha}{1 + \tan^2 \alpha}\right)} \tag{5.85}$$

For $\alpha = 0$ and $\alpha = \pi/2$, the eqn. above gives $\lambda_l = \lambda_x$ and $\lambda_l = \lambda_y$ respectively, and thus for $\alpha = 0$ and $\alpha = \pi/2$, the eqn. (5.85) will automatically become converted into its special case (5.80) and (5.83) respectively. If $n = 1$, i.e. if $\lambda_x = \lambda_y$, then (5.85) gives $\lambda_l = \lambda_x$, i.e. $\lambda_l = \lambda_x = \lambda_y$.

It follows that in a distorted model, the Froudian velocity scale can, in fact, be the scale of the velocities *in the direction x of the flow* only. For example, it can be the scale of, say, average or maximum velocity of the river or channel flow, of the speed of propagation of a swell, or of at flood wave, and so on, but it cannot be, for example, the scale of the velocity of rise and fall of the water level during passage of the flood wave, for the latter velocity is in the direction y, and thus its scale is determined by (5.83). Observe that the scales of the velocities U_x and U_y are related to each other as follows

$$\frac{\lambda_{u_y}}{\lambda_{u_x}} = n = \text{distortion} \tag{5.86}$$

REFERENCES

1. V. J. Galay, *Observed Forms of Bed Roughness in an Unstable Gravel River*, XII Congress of the I.A.H.R., Fort Collins, Colorado (1967).
2. G. H. Keulegan, 'Laws of Turbulent Flow in Open Channels', U.S. National Bureau of Standards, *Jour. Res.*, V, 21, Paper No. 1151 (1938).
3. H. Basin, Recherches hydrauliques, Mém. divers savants, *Sci. Math. et Phys.*, **19** (1865).
4. Ven Te Chow, *Open Channel Hydraulics*, McGraw-Hill (1959).
5. R. C. H. Russell, 'Methods of Selecting Scales for Models in use atthe Hydraulics Research Station (Wallingford)', Report No. INT. 40, Wallingford (1964).
6. M. S. Yalin, *Scales of River and Channel Models*, Ch. 28 in *River Engineering and Water Conservation Works* (Ed., R. B. Thorn), Butterworths (1966).
7. H. Schlichting, *Boundary Layer Theory*, McGraw-Hill (1968).

6 Similarity in Sediment Transport

6.1 General Criteria of Similarity

Consider the flow in a channel consisting of erodible granular material. The shear stress τ_0 acting on the flow boundary varies along the wetted perimeter in a manner shown schematically by the τ_0-diagram in Fig. 6.1.

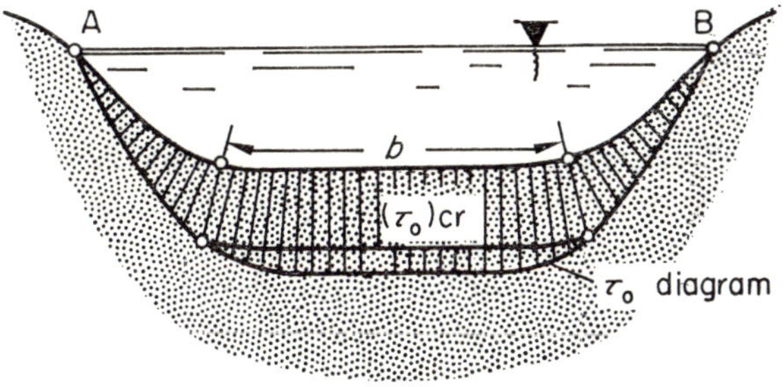

FIG. 6.1

The value of τ_0 is equal to zero at the points A and B (at the free surface); the largest values of τ_0 being at the lowest part of the wetted perimeter (at the bed). The magnitude of the τ_0 diagram increases with the intensity (average velocity v) of the flow. Since the grains forming the flow boundary possess a finite weight in the fluid and a finite coefficient of friction, they can be moved by the shear stress τ_0, only if it exceeds a certain finite value denoted by $(\tau_0)_{cr}$ and referred to as *critical shear stress*. The value of $(\tau_0)_{cr}$ is dependent on the nature of the granular material and on the nature of the fluid. If the values of τ_0 corresponding to any point on the flow boundary is less than $(\tau_0)_{cr}$, then the granular material is not moved and the flow boundary is virtually a 'rigid boundary' (studied in the preceding chapter). If, on the other hand, the intensity of the flow (and thus the magnitude of the τ_0-diagram) is sufficiently high so as to yield a region *b* where

$$\tau_o > (\tau_o)_{cr}$$

is valid, then the grains of the region *b* will be moved. In such a case, referring to the region *b* as the *flow bed*, it is said that the flow is transporting the bed material (sediment) or that sediment transport is taking place. Clearly, the motion of the given bed material depends entirely on the mechanical character of the flow which generates it. On the other hand, since the motion of the bed material is accompanied by certain features of its own (such as the diffusion of solid particles into the liquid medium, the wave-like deformation of the surface of the bed (sand waves) etc.), and since these features affect the motion of the fluid to a considerable extent, it follows that the motion of the transported sediment and the motion of the transporting fluid are interdependent. Hence, the simultaneous motion of two phases (liquid and solid) constitute an inseparable mechanical totality, which will be referred to as the *two phase motion* (or phenomenon).

Let us first consider steady and uniform two phase motion, and let us attempt to establish the set of characteristic parameters of this motion. The geometry of the two phase motion under consideration is completely determined by:

(i) the shape (geometry) of the cross-section of the flow. (It is assumed that the channel possesses a trapezoidal shape, as in Fig. 5.1, with a non-vanishing central region B_c. This is the only restriction imposed on the geometry of the cross-section.)

(ii) the geometric properties of the granular material (i.e. the properties considered in Section 4.1).

In the study of filtration (Chapter 4) it was possible to introduce the concept of the effective diameter, that is, it was possible to determine such an absolute size $\bar{D}$ of the uniform material, consisting of spherical grains, that could yield (approx.) the same values for the property Π_J as the given granular material. In the case of sediment transport, we are not dealing with only one property (such as J or Π_J in filtration), we are interested in the behaviour of various properties of the two phase motion (transport rate, grain velocity, distribution of concentration, sand wave length, friction factor, etc.). There is no reason to assume that a particular size of the uniform material which would yield approximately the same values for dimensionless transport rate (throughout the stages of the flow) will also yield (throughout the stages) the same values for dimensionless sand wave length, sand wave height, friction factor and so on. On the contrary, it must be expected that every different dimensionless property Π_A will require its own different effective diameter. It follows that, in the case of two phase motion, the concept of effective diameter cannot be adopted, if only because the effective diameter can be 'effective' with respect to only one property; it cannot be equally effective for all properties of the two phase motion. One is often confronted by the objection 'surely there must exist a uniform material of a particular size $\bar{D}$ that moves in exactly the same manner as the given material . . .'. The invalidity of this objection lies in the interpretation of motion as a single property. The motion (of granular material) is merely the name given to the totality of many properties; e.g. $A_1 =$ grain velocity, $A_2 =$ grain acceleration, $A_3 =$ length of saltation, $A_4 =$ number of detachments (per unit time) . . . etc. Hence, 'moves in the same manner' cannot be interpreted, e.g. as $A \equiv \bar{A}$; it can be interpreted only as an unlimited set of identities $A_i \equiv \bar{A}_i$ $(i = 1, 2, \ldots \infty)$. But this brings us precisely to what was meant above when it was stated that 'we are interested in the behaviour of *various* properties'.

From these explanations, it follows that when dealing with the reproduction of two phase motion in a model, the safest approach would be to use model granular material that is geometrically similar to that of the prototype. Accordingly, in the following study we will invariably assume that the model granular material is selected (or prepared) so as to have the shape of the grains (statistically speaking) and the shape of the dimensionless grain size distribution curve, at least approximately the same as those of the prototype. This being so, the model and prototype grain sizes D_i'' and D_i', corresponding to any common fraction i, will be related to each other by the same constant proportion

$$\frac{D_i''}{D_i'} = \lambda_D \qquad \text{(for any } i\text{)}$$

When determining the absolute size of a granular material the unit used for the measurement must be indicated. In such cases, we will depart from the usual symbol D representing *any* typical diameter, and will mark the particular grain size used as the unit with a corresponding subscript. To maintain consistency in the method the absolute size of granular material will always be 'measured' using D_{50} as the unit. Hence, the size of the prototype material (having a specified geometry) will be given for example, as $D_{50}' = 3{\cdot}5$ mm.

The roughness k_s of a plain granular bed can be expressed as follows

$$k_s = \alpha D_{50}$$

where the proportionality factor α, strictly speaking, varies with the geometry of the granular material (with the shape of the grains and the grain size distribution curve). However, the form of this variation has not yet been revealed (i.e. at present, we do not know what are the values of α if the grain size distribution curve is of convex or concave type; if the grains are 'round' or 'flat', etc.). Hence, in the following considerations we will have no alternative but to assume, as is usually done, that $\alpha \approx 1$, i.e. that the approximate equality

$$k_s = D_{50}$$

is valid for any geometry of granular material. In other words, we will assume that with respect to the particular property $A = k_s$, the effective diameter is D_{50}. Fortunately, this approximate equality will be used mainly (in connection with $v_* k_s/\nu$) for determining the limits of some regions of validity, not for computing the scales themselves.

In the present chapter, only cohesionless granular material will be considered. Furthermore, it will be assumed (in accordance with practice) that the grains composing the granular material have the same density ρ_s, and thus the same specific weight, in air

$$\bar{\gamma}_s = \rho_s g \tag{6.1}$$

and the same specific weight in fluid (having the density ρ).

$$\gamma_s = \bar{\gamma}_s - \rho g = g(\rho_s - \rho) \tag{6.2}$$

Provided the assumptions above are observed, two phase phenomenon is completely determined by:

(i) the nature of the fluid (i.e. by its viscosity μ and density ρ),
(ii) the nature of the cohesionless granular material (of specified geometry) which can be completely described by its absolute size, i.e. by its typical diameter D, and by the grain density ρ_s, and
(iii) the nature of the uniform flow, that is, by the slope S, by the absolute size (flow depth) h of the cross-section (having a certain specified geometry), and by the acceleration due to gravity g which brings the flow, and consequently the whole two phase motion into being.

Hence, the steady and uniform two phase motion under consideration can be described by the following set of seven independent quantities (characteristic parameters)

$$\mu, \rho, D, \rho_s, S, h, g \tag{6.3}$$

5 Aberdeen Channel model, Hydraulics Research Station, Wallingford

6 Telek Anson model, Hydraulics Research Station, Wallingford

Observe that the consideration of the parameters S and g can be substituted by the following combinations of the characteristic parameters which include S and g

$$\left.\begin{aligned} v_* &= \sqrt{(gSh)} \\ \gamma_s &= g(\rho_s - \rho) \end{aligned}\right\} \tag{6.4}$$

Accordingly, the consideration of the set (6.3) can be replaced by that of

$$\mu,\ \rho,\ D,\ \rho_s,\ v_*,\ h,\ \gamma_s \tag{6.5}$$

As is clear from (6.4) the parameter combination γ_s is the specific weight of grains in fluid, while v_* is the shear velocity formed by the shear stress $\tau_0 = \rho g S h$ acting on the channel bed (b) of a trapezoidal channel (possessing a sufficiently large value of the width-depth ratio). Hence the set (6.5), which contains as parameters the combinations γ_s and v_*, into which the quantities S and g must necessarily enter when appearing in any mathematical expression of the two phase motion, is, from a physical point of view, more significant than its mathematical equivalent (6.3). In the following, preference will always be given to the version (6.5).

The sets (6.3) and (6.5) have been determined assuming that the fluid entering the channel is *clear*, i.e. that it does not possess any *initial* amount C_o of solid particles in suspension ($C_o \equiv 0$). Certainly, the fluid entering the channel without any initial concentration or 'charge' might acquire a certain suspended load as the result of its motion *in* the channel. But in this case, the amount of particles transported in suspension (as any other property of the phenomenon) cannot be anything else but a certain function of the characteristic parameters (6.5). On the other hand, it is always possible that the fluid entering the channel is not clear and may possess an initial charge C_o. Since such an initial concentration C_o has been acquired by the fluid before entering the channel, it can be of any physically possible value, and thus it has to be regarded as an independently varying quantity (having a certain influence on the subsequent destiny of the two phase motion in the channel). It follows that, if an initial concentration C_o is present, then it must be considered as an additional characteristic parameter, in which case the number of characteristic parameters will rise to $n = 8$. Accordingly, the present set (6.5) can be regarded as the special case of the mentioned set of $n = 8$ variables (which corresponds to the special value $C_o = 0$).*

* The author recalls with appreciation his discussion with R. C. Muddock during the XII Congress of I.A.H.R. (1967, Fort Collins) which led to these clarifications.

Let us now go over to determine the dimensionless combinations that follow from the set (6.5). Selecting ρ, v_* and D as basic quantities, we arrive at the following $n - 3 = 7 - 3 = 4$ dimensionless variables

$$\left.\begin{aligned} X_1 &= \rho v_* D \mu^{-1} = \frac{v_* D}{\nu} \\ X_2 &= \rho v_*^2 D^{-1} \gamma_s^{-1} = \frac{\rho v_*^2}{\gamma_s D} \\ X_3 &= \rho^0 v_*^0 D^{-1} h = \frac{h}{D} \\ X_4 &= \rho^{-1} v_*^0 D^0 \rho_s = \frac{\rho_s}{\rho} \end{aligned}\right\} \tag{6.6}$$

and at the following dimensionless version of a property A

$$\Pi_A = \rho^{x_A} v_*^{y_A} D^{z_A} A \tag{6.7}$$

e.g. if A is the volumetric flow rate Q, then

$$\Pi_Q = \rho^0 v_*^{-1} D^{-2} Q = \frac{Q}{v_* D^2}$$

if A is the (specific) transport rate p of sediment (weight per unit width and time), then

$$\Pi_p = \rho^{-1} v_*^{-3} D^0 p = \frac{p}{\rho v_*^3}$$

etc.

Clearly, dynamic similarity of the two phase motion is given by the conditions

$$\lambda_{X_1} = 1;\ \lambda_{X_2} = 1;\ \lambda_{X_3} = 1;\ \lambda_{X_4} = 1 \tag{6.8}$$

* The dimensionless combinations (variables) (6.6) are well-known in the theory of sediment transport. Among them X_1 and X_2 were apparently first introduced by A. Shields[1] in 1936. The combination Θ of R. A. Bagnold[2] (1956), and the parameter Ψ^{-1} of H. A. Einstein[3,4,5] (1942, 1950 and 1965) are, in principle, identical to the combination $X_2 = \tau_0/\gamma_s D$ of A. Shields. The ratio X_3 was systematically taken into account for the study of the transport rate by J. Rottner[6] (1959). The simultaneous consideration of all four dimensionless combinations X_1, X_2, X_3, and X_4 as determining variables of any dimensionless quantity related to the two phase motion is due to M. S. Yalin[7,8,9] (1963, 1964 and 1965).

i.e. by the scale relations

$$\left.\begin{aligned} \lambda_{v_*}\lambda_D &= \lambda_\nu \\ \lambda_{v_*}^{2}\lambda_{\gamma_s}^{-1}\lambda_D^{-1} &= \lambda_\rho^{-1} \\ \lambda_h\lambda_D^{-1} &= 1 \\ \lambda_{\rho_s} &= \lambda_\rho \end{aligned}\right\} \quad \text{or} \quad \left.\begin{aligned} \lambda_{v_*} &= \lambda_\nu^{\frac{1}{3}} \\ \lambda_h &= \lambda_\nu^{\frac{2}{3}} \\ \lambda_D &= \lambda_\nu^{\frac{2}{3}} \\ \lambda_{\rho_s} &= \lambda_\rho \\ \lambda_{\gamma_s} &= \lambda_\rho \end{aligned}\right\}^{*} \qquad (6.9)$$

From the second set in (6.9) it is clear that only two scales (e.g. γ_ν and λ_ρ) can be selected freely. This is because the scale λ_g has already been selected as unity when deriving the second set from the first (see the footnote). It follows that if λ_ν and λ_ρ are chosen as unity, i.e. if the model operates with the prototype fluid (water), then the dynamically similar reproduction of sediment transport in a small scale model is a theoretically predictable impossibility. At the same time, it also follows that all the well-known difficulties in reproducing sediment transport in a small scale model occur not because we do not know what the criteria of similarity are, but because we cannot apply them under existing technical and economical restrictions, which compel us to use water in the model. To put it differently, the difficulty arises because Nature could not forsee when creating her laws, that we would be willing to vary the values of the parameters D, h, v_*, γ_s, and ρ_s, but would be reluctant to vary ρ and μ!

Since the scale relations (6.9) follow directly (without any additional assumptions) from the set of characteristic parameters (6.5), and since under the conditions introduced, the sufficiency of the set (6.5) is clear to anyone familiar with the theory of dimensions, there is hardly any need for experimental evidence to convince one of the correctness of the scale relations (6.9). On the other hand, a visual demonstration of the possibility of dynamically similar reproduction of sediment transport in a small scale model, irrespective of whether such a model can or cannot be used in practice, is undoubtedly desirable. Accordingly, at the Hydraulics Research Station, Wallingford (where the scale relations (6.9), were developed) the similarity flumes shown in Plate 2b were constructed. The larger and smaller flumes (prototype and model) were prepared strictly observing the conditions of geometric similarity (the similarity of guide vanes and pumps being included). The scale relations (6.9) were satisfied by having:

* One arrives from the first set of (6.9) to the second by taking into account that $\gamma_s = g(\rho_s - \rho)$, and thus that $\lambda_{\gamma_s} = \lambda_g\lambda_\rho[(X_4'' - 1)/(X_4' - 1)]$. The validity of $\lambda_{X_4} = 1$, i.e. of $X_4'' = X_4'$ implies that the multiplier in the square brackets is unity. In addition $\lambda_g = 1$. Hence $\lambda_{\gamma_s} = \lambda_\rho$.

Table 6.1

	Prototype	Model
Fluid	$\mu' = 5{\cdot}89$, $\rho' = 1{\cdot}125$, $\nu' = 5{\cdot}24$ (49% solution of glycerol in water)	$\mu'' = 1{\cdot}00$, $\rho'' = 1{\cdot}00$, $\nu'' = 1{\cdot}00$ (water)
Granular material	$\gamma_s' = 1{\cdot}26$; $D' = 0{\cdot}96$ mm (coal)	$\gamma_s'' = 1{\cdot}12$; $D'' = 0{\cdot}32$ mm (araldite epoxy resin)

and thus

$$\lambda_h = \lambda_D = \tfrac{1}{3};\ \lambda_\nu = 0{\cdot}192;\ \lambda_{\gamma_s} = \lambda_{\rho_s} = \lambda_\rho = 0{\cdot}887$$

The necessity of having geometrically similar grain size distribution curves was eliminated by having uniform (D = const.) granular materials. Geometric similarity of the shape of the particles could not have been achieved, even approximately. The required constant ratio $\lambda_Q = \lambda_h^{2\cdot5}$ (see the end of the present section) between the flow rates was obtained by driving the two pumps from a common variable speed drive. From Plate 2b one can see the almost perfect similarity of two features (waves) formed on the coal and resin beds during the corresponding two stages.

Since $\lambda_\nu = \lambda_\rho = 1$ yields unity for all the remaining scales in equation (6.9), the design of a practical model according to the model and prototype identity of all four dimensionless variables X_1, X_2, X_3 and X_4 is impossible. There is no alternative therefore to seeking special solutions where the identity of some of the dimensionless variables can be relaxed, and so make small-scale models a practical possibility.

6.2 Large Values of the Reynolds Number X_1

If the Reynolds number X_1 is larger than a certain critical value,** then the

* The properties of model fluid (water at 20°C) are taken as *units*; the properties of the prototype being given in terms of these units. Accordingly γ_s' and γ_s'', in fact, are the ratios γ_s'/γ' and γ_s''/γ'' respectively (where $\gamma'' = \rho''g$ and $\gamma' = \rho'g$). Hence there are no dimensions in the table for μ, ρ, ν and γ_s.

** Even if the typical grain size D is defined (i.e. even if it is specified as, say, D_{50}) the critical value of X_1 must be dependent on the geometry of the flow and granular material. Hence, one cannot give a definite value of the critical X_1 that would be valid for all types of flow cross-section and for all grain size distribution curves and shapes of the grains. However, the experimental data indicates that the critical values of X_1 should be somewhere in the range $\approx 70 < X_1 < \approx 150$ (see e.g. the Shields curve in Fig. 6.2).

viscosity μ, and thus the Reynolds number X_1 itself, can no longer be a variable with regard to the detachment and motion of the grains. In this case the first condition of the first set of (6.9) (which follows from $\lambda_{X_1} \equiv 1$) can be excluded from consideration. By so doing, and considering that $\lambda_{\gamma_s} = \lambda_{\rho_s} = \lambda_\rho$ we reduce (6.9) into

$$\left.\begin{aligned} \lambda_{v_*} &= \sqrt{(\lambda_h)} \\ \lambda_h &= \lambda_D \\ \lambda_{\rho_s} &= 1 \end{aligned}\right\} \qquad (6.10)$$

The bed material of the model implied by (6.10) consists of grains reduced in the same proportion as the external dimensions of the flow (i.e. the model is geometrically similar, including granular material), while the density or the specific weight of the model and prototype grains are identical, i.e. the model bed material is formed by sand or gravel. Let us combine the scales (6.10) with the scales (5.9) corresponding to the geometrically, and dynamically, similar model with a rough turbulent flow. In other words, let us assume that the dynamically similar model studied in 5.2.2 has a mobile bed. The relations (5.9) are

$$\left.\begin{aligned} \lambda_v &= \sqrt{(\lambda_l)} \\ \lambda_l &= \lambda_{k_s} \\ \lambda_S &= \lambda_\rho = 1 \end{aligned}\right\} \qquad (6.11)$$

The exclusion of X_1 implies that X_1', and thus D_{50}' have to be regarded as large, and consequently that the sand waves are either underdeveloped or they are not present at all. But if so, then the size k_s of the roughness of the bed must be proportional to say D_{50} (see the Section 6.1) which implies that

$$\lambda_{k_s} = \lambda_{D_{50}} = \lambda_D$$

must be valid. Furthermore λ_h in (6.10) and λ_l in (6.11) have the same meaning. Hence, the sets (6.10) and (6.11) can be unified into

$$\left.\begin{aligned} \lambda_v &= \lambda_{v_*} = \sqrt{(\lambda_l)} \\ \lambda_l &= \lambda_D = \lambda_{k_s} \\ \lambda_S &= \lambda_\rho = \lambda_{\rho_s} = 1 \end{aligned}\right\} \qquad (6.12)$$

The scale of any property A is given by

$$\lambda_A = \lambda_\rho^{-x_A} \lambda_{v_*}^{-y_A} \lambda_D^{-z_A}$$

(see (6.7)), and taking into account (6.12) by

$$\lambda_A = \lambda_l^{-\frac{y_A}{2}} \lambda_l^{-z_A} = \lambda_l^{-\left(\frac{y_A}{2} + z_A\right)} \qquad (6.13)$$

e.g. if A is the flow rate Q, then $y_A = -1$; $z_A = -2$, and thus

$$\lambda_Q = \lambda_l^{2\cdot5}$$

if A is the sediment transport rate p, then $y_A = -3$; $z_A = 0$, and thus

$$\lambda_p = \lambda_l^{1\cdot5}$$

etc.

EXAMPLE 6.1

Consider the prototype given by

$$h' = 3{\cdot}00 \text{ m};\ S' = 0{\cdot}004;\ D_{50}' = 20 \text{ mm};\ \gamma_s'/\gamma = 1{\cdot}65$$
$$(g = 9{\cdot}81 \text{ m/s}^2;\ \nu = 10^{-6} \text{ m}^2/\text{s}\ (\approx 20°\text{C}))$$

It is intended to establish whether, at the flow stage given by the characteristics above, the transport of granular material is present, and if the

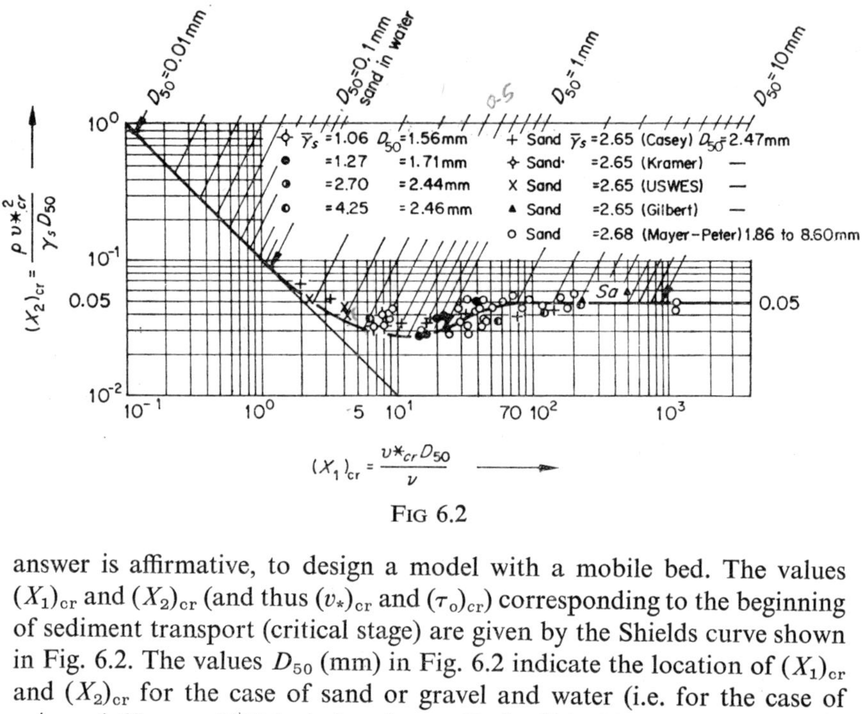

FIG 6.2

answer is affirmative, to design a model with a mobile bed. The values $(X_1)_{cr}$ and $(X_2)_{cr}$ (and thus $(v_*)_{cr}$ and $(\tau_o)_{cr}$) corresponding to the beginning of sediment transport (critical stage) are given by the Shields curve shown in Fig. 6.2. The values D_{50} (mm) in Fig. 6.2 indicate the location of $(X_1)_{cr}$ and $(X_2)_{cr}$ for the case of sand or gravel and water (i.e. for the case of $\gamma_s/\gamma = 1{\cdot}65$, $\gamma = 1000$ kg/m³ and $\nu = 10^{-6}$ m²/s). The prototype values of the variables X_1 and X_2 are

$$X_1' = \frac{v_*' D_{50}'}{\nu} = \frac{\sqrt{(9{\cdot}81 \times 0{\cdot}004 \times 3{\cdot}00)}0{\cdot}02}{10^{-6}} = 6860^*$$

* See the text just after the eqns. (5.17), (5.18) and (5.19).

and

$$X_2' = \frac{\rho v_*'^2}{\gamma_s' D_{50}'} = \frac{\gamma}{\gamma_s'} \frac{S'h'}{D_{50}'} = \frac{1}{1{\cdot}65} \frac{0{\cdot}004 \times 3{\cdot}00}{0{\cdot}02} = 0{\cdot}364$$

Hence, the flow is rough turbulent and sediment transport is present, for the value of X_2' is much larger than $0{\cdot}05 = (X_2')_{cr}$ corresponding to all large grain sizes (see Fig. 6.2). The model scale λ_l must satisfy the condition

$$\lambda_l > \left(\frac{70}{X_1'}\right)^{\frac{2}{3}} = \left(\frac{70}{6860}\right)^{\frac{2}{3}} = \frac{1}{21{\cdot}25}$$

which is simply (5.16) where $k_s \approx D_{50}$ is substituted. Selecting

$$\lambda_l = \frac{1}{20}$$

and using this value in the relations (6.12), we obtain

$$\lambda_v = \lambda_{v*} = \frac{1}{\sqrt{20}} = \frac{1}{4{\cdot}475}$$

and

$$\lambda_D = \frac{1}{20}$$

Hence, the model characteristics are

$$\begin{aligned} S'' &= 0{\cdot}004 \\ D_{50}'' &= 2/20 = 0{\cdot}1 \text{ cm} = 1 \text{ mm} \\ \gamma_s'' &= 1{\cdot}65 \text{ ton/m}^3 \\ h'' &= 300/20 = 15 \text{ cm} \end{aligned} \quad \left.\begin{matrix} \\ \\ \end{matrix}\right\} \text{sand}$$

In accordance with the explanations given at the beginning of the chapter, it is assumed that the model and prototype granular materials are geometrically similar, i.e. that the model sand is selected (or prepared) so as to have approximately the same shape of grains and of dimensionless grain size distribution curve as the prototype granular material.

As the model value of the Reynolds number $v_* k_s/\nu$, we determine

$$\frac{v_*'' k_s''}{\nu} \approx \frac{v_*'' D_{50}''}{\nu} = X_1'' = \lambda_{v_*} \lambda_D X_1' = \frac{1}{4{\cdot}475} \times \frac{1}{20} \times 6860 = 72{\cdot}2$$

which just exceeds ≈ 70 (because the scale $\lambda_l = 1/20$ just exceeds the lower limit $1/21{\cdot}25$). Since the eqn. (6.13) is valid for the scale of any property A, we determine

$$\lambda_A = \left(\frac{1}{20}\right)^{-\left(\frac{y_A}{2} + A_Z\right)}$$

Hence, if A is the flow rate Q, then

$$\lambda_Q = \frac{1}{20^{2\cdot5}} = \frac{1}{1790}$$

if A is the sediment transport rate p, then

$$\lambda_p = \frac{1}{20^{1\cdot5}} = \frac{1}{89\cdot5}$$

6.3 Non-uniform and Non-stationary Two Phase Phenomena

With respect to the combination (6.12) of the two sets (6.10) and (6.11), the following question can be raised: The set (6.10) corresponds to a uniform flow, and the set (6.11) to a non-uniform flow. Is it permissible to combine them? Let us attempt to answer this question in a general manner, i.e. by considering all four variables X_j and by taking into account the possibility of the variation of the flow, and thus of the two phase phenomenon, with time as well as with location.

Consider how the expression of a section property A corresponding to a steady and uniform two phase phenomenon differs from the expression of the same property corresponding to a non-steady and non-uniform phenomenon. In the case of a steady and uniform phenomenon we have

$$\Pi_A = \varphi_A(X_1, X_2, X_3, X_4) \tag{6.14}$$

If, however, the phenomenon is neither steady nor uniform, then the expression of Π_A must be generalized into

$$\Pi_A = \varphi_A\left(X_{1_0}, X_{2_0}, X_{3_0}, X_4, \frac{\partial\psi_1}{\partial\xi}, \frac{\partial\psi_2}{\partial\xi}, \frac{\partial\psi_1}{\partial\theta}, \frac{\partial\psi_2}{\partial\theta}, \xi, \theta\right) \tag{6.15}$$

Here ξ and θ stand for the dimensionless abscissa and time respectively, while ψ_1 and ψ_2 are the dimensionless functions representing the variations of v_* and h

$$\frac{v_*}{(v_*)_0} = \psi_1(\xi, \theta) \tag{6.16}$$

$$\frac{h}{h_0} = \psi_2(\xi, \theta)$$

$(v_*)_0$ and h_0 being the typical values of v_* and h corresponding to the typical section $\xi = \xi_0$ and typical time $\theta = \theta_0$. The subscript $_0$ in the first three X_{j0} indicates that they are formed by $(v_*)_0$ and h_0.

If the model is geometrically similar, as the undistorted model studied in the preceding section, then the functions ψ_1 and ψ_2 have identical forms in model and prototype. In this case, when comparing the model and prototype values Π_A for the corresponding sections and times (for the identical values of ξ and θ), we will always have the identity of the dimensionless quantities that appear in (6.15) as the functions of ξ and/or θ. Hence, the identity of model and prototype values of each Π_A will be given again by the identity of the dimensionless combinations X_j alone.

If, however, the model is distorted (like that studied in the next subsection) then one would expect that the form of the functions ψ_1 and ψ_2 would be affected by the distortion to some extent. In this case, we have the following alternatives:

(i) to restrict the consideration of similarity to the central regions only, where the flow and the two phase phenomenon can be regarded as two-dimensional, (for the two-dimensional flows can always be treated as geometrically similar)

(ii) to ignore the influence of the derivatives

$$\frac{\partial \psi_1}{\partial \xi}, \frac{\partial \psi_2}{\partial \xi}, \frac{\partial \psi_1}{\partial \theta}, \frac{\partial \psi_2}{\partial \theta}$$

altogether on the grounds that, in natural open channels and rivers, the variation of v_* and h, with distance and time, can hardly be so drastic that the influence of the rates of change of v_* and h (on a property A) can become comparable with the influence of v_* and h themselves. So far not a single relation for any property A of two phase motion (transport rate, sand wave length, friction factor, etc.) has been produced that contains the derivatives of v_* or h (with respect to distance or time) apart from these parameters themselves. Accordingly, any analysis that is carried out today for non-steady and non-uniform two phase phenomena is, in fact, carried out by means of using steady and uniform flow formulae, for transport rate, friction factor, etc. In other words, contemporary analysis rests on the assumption that a non-steady and non-uniform phenomenon can be divided (with regard to distance and time) into small segments δx and δt where the flow can be treated as steady and uniform, i.e. where the influence of both convective and temporal accelerations can be regarded as negligible. Clearly, such a method of study is precisely what is meant by the alternative (ii). This method is certainly acceptable as far as natural river and channel flows are concerned; however, it cannot be regarded, as acceptable in general (e.g. sediment transport by short wind waves).

6.4 The Investigation of Scour

The design method implied by (6.12) also covers the investigation of *scour* that can take place on the downstream side of a hydraulic structure. For a given flow rate, the mechanical character of the flow past a hydraulic structure (weir, spillway of a dam, etc.) is determined entirely by the geometry of that structure and its environment. Hence, the investigation of the flow past a structure and its immediate environment has to be carried out in a geometrically similar (undistorted) model such as is shown in Plate 3. But the only possible way for the realization of an undistorted small scale model involving a mobile bed is that implied by (6.12). Accordingly, a reliable investigation of the scour can be carried out only if the model can be designed in accordance with (6.12). As is clear from Section 6.2, the bed material of such a model should be sand or gravel, and the flow rough turbulent. Confining the following considerations to the two-dimensional regions, and taking into account the validity of $\lambda_l = \lambda_D = \lambda_{k_s}$, the condition (5.16) for the existence of rough turbulent flow in the model can be expressed as follows:

$$\lambda_D > \left(\frac{70}{X_1'}\right)^{\frac{2}{3}} \tag{6.17}$$

where

$$X_1' = \frac{v_*' D_{50}'}{\nu}$$

Unlike the flow in a long open channel, the determination of v_*' for various sections of the flow past a structure will be far more reliable if it is expressed by the average velocity v' corresponding to those sections. Thus we can write

$$v_*' = \frac{v'}{c'} \tag{6.18}$$

which gives as (6.17)

$$\lambda_D > \left(\frac{70c'}{\dfrac{v' D_{50}'}{\nu}}\right)^{\frac{2}{3}} \tag{6.19}$$

Assuming that in the region immediately downstream of the structure the velocity distribution is logarithmic, and taking into account that we are dealing with a rough turbulent flow, we can express v' and c' in the following manner

$$v' = \frac{q'}{h'} \tag{6.20}$$

$$c' \approx 2{\cdot}5 \ln\left(11 \frac{h'}{D_{50}'}\right)^{*} \tag{6.21}$$

where q' is the flow rate per unit width of the structure. Using these values in (6.19) we arrive at

$$\lambda_l = \lambda_D > \left[\left(175 \frac{v'}{q'}\right) X_3' \ln(11X_3')\right]^{\frac{2}{3}} \tag{6.22}$$

i.e.

$$D_{50}'' > \left[h' \left(175 \frac{v'}{q'}\right) \ln(11X_3')\right]^{\frac{2}{3}} [D_{50}']^{\frac{1}{3}} \tag{6.23}$$

where X_3' implies h'/D_{50}'.

EXAMPLE 6.2

Consider the weir shown in Fig. 6.3. The flow rate per unit width is $q' = 10\ \mathrm{m^2/s}$, the weir height $Z' = 10$ m; the absolute size of the cohesionless bed material (having a specified geometry) being given by $D_{50}' = 3{\cdot}5$ cm.

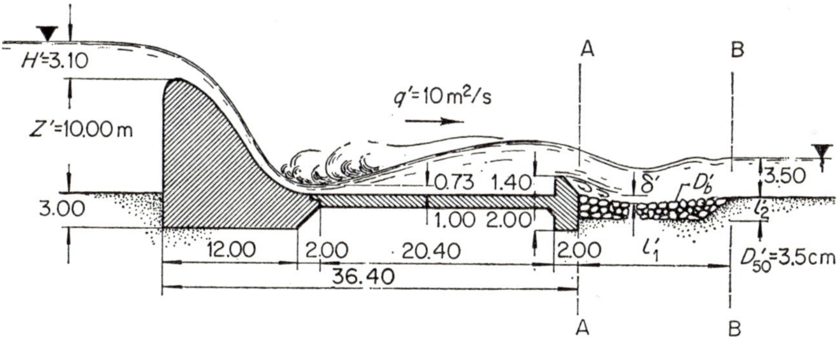

FIG 6.3

It is intended to design the scour bed (adjacent to the end sill of the energy dissipator), i.e. to determine

(i) the shape,
(ii) the size (dimensions l_1', l_2'), and
(iii) the composition (size D_b' of the boulders)

* Here again we assume that the size k_0 of the roughness of the plain mobile bed can be approximated by the size D_{50} of the mobile bed material.

of the protective filling (riprap) so that the depth δ' of erosion does not exceed a certain specified value during the passage of $q' = 10\ m^2/s$ which is assumed to be the design flow rate. Determine the scales of an undistorted model where various versions of (i), (ii) and (iii) can be tested and the most appropriate solution can be found.

If the flow at the section B—B (where the roughness and velocity are smallest) is rough turbulent, then it must be such throughout the region between the sections A—A and B—B. Accordingly, we express the condition (6.22) by the properties of the section B—B. Substituting the value

$$X_3' = \frac{3{\cdot}50}{0{\cdot}035} = 100$$

corresponding to the section B—B in (6.22) we obtain

$$\lambda_l > \left[\frac{175 \times 10^{-6}}{10} \times 10^2 \ln (1100)\right]^{\frac{2}{3}}$$

i.e.

$$\lambda_l > \left[\frac{1}{81{\cdot}5}\right]^{\frac{2}{3}} = \frac{1}{18{\cdot}8}$$

Hence, the condition (6.22) does not present any difficulty, for if the height of the model weir is, say, $Z'' = 60$ cm, then the model scale is

$$\lambda_l = \frac{Z''}{Z'} = \frac{60}{1000} = \frac{1}{16{\cdot}67}$$

and thus the inequality above is satisfied with a reasonable margin. Adopting $\lambda_l = 1/16{\cdot}7$, we determine the following value for the size of the (geometrically similar) model bed material (sand)

$$D_{50}'' = \frac{35}{16{\cdot}7} = 2{\cdot}1 \text{ mm}$$

Clearly, the simplicity of this solution is due to the fact that the size of the prototype bed material was as large as $D_{50}' = 3{\cdot}5$ cm. What is the *minimum* value of D_{50}' that would still permit the prototype described above to be tested in an undistorted model of the scale λ_l? Using the equation

$$\lambda_l = [175 \times 10^{-7} \times X_3' \times \ln 11 X_3']^{\frac{2}{3}}$$

we determine for various λ_l corresponding values of X_3' and subsequently compute the prototype grain size D_{50}'. The results of such computations are shown in the table below (the heights Z'' of the model weirs are also included).

Table 6.2

λ_l	$\frac{1}{16{\cdot}7}$	$\frac{1}{11{\cdot}2}$	$\frac{1}{8{\cdot}2}$	$\frac{1}{6{\cdot}65}$	$\frac{1}{5{\cdot}7}$
Z'' (cm)	60	89	120	150	175
X_3'	100	200	300	400	500
D_{50}' (cm)	3·5	2·35	1·72	1·40	1·20

From this table, which is valid for $q' = 10\ \mathrm{m^2/s}$ one can see clearly how the model dimensions increase when the prototype grain size decreases. One can say that the design method demonstrated in the present example can be used if the model height Z'' it yields does not exceed, say 70 to 80 cm (the value of Z'' being dependent, of course, on both q' and D'). The two-dimensional weir model having the height say $Z'' = 75$ cm (and length of the order of 2 to 3 m) should not be regarded as unduly large. If a distance as long as 4 or 5 weir lengths is allowed for the flow on the downstream side of the weir and 1 or 2 weir lengths for the flow on the upstream side, the total length of the two-dimensional model would be ≈ 20 m. Such a model could be conveniently accommodated in a, say, 20 m long flume having a cross-section, say, 1×1 m. Flumes of this size are part of the equipment of many hydraulic laboratories.

The crest lengths of the weirs are usually many times longer than the maximum value of the weir load H. Accordingly, the flow past the central region of a weir can usually be treated as a two-dimensional flow. Often it is preferable to carry out the investigations related to scour for a two-dimensional flow corresponding to the central region of the weir, rather than attempting to achieve more natural conditions for the flow geometry by using a smaller three-dimensional model (which would be bound to have the mobile bed formed by an incorrect granular material). The deviation from actual three-dimensional conditions may produce only a few percent of error in the longitudinal section of the scour corresponding to the central part of the weir; on the other hand, the deviation from the correct granular material might cause errors of several hundreds percent.

6.5 Transport of Bed Material en masse

If the prototype grain size D_{50}', and thus the order of the Reynolds number X_1' is small ($X_1' < \approx 70$), then the influence of X_1' cannot be neglected in so far as it affects both sediment transport and its initiation. Accordingly, if X_1' is small, then the first equation of (6.9) which follows from the condition $\lambda_{X_1} = 1$ cannot be neglected, and a different way has to be found for the reproduction of two phase motion in a small scale model.

Even though the dimensionless formulation of any property A can be ensured only if it is treated as a function of four variables

$$\Pi_A = \varphi_A(X_1, X_2, X_3, X_4) \tag{6.24}$$

The importance of each variable in the functional relation above varies, depending on the order of the variable itself, as well as on the nature of the property A (e.g. if the Reynolds number X_1 is large, then its influence becomes negligible). For example, if A is a property related to the motion of an individual grain, then the ballistics of the accelerated motion of a (bed load) grain by *saltation* must certainly be dependent on its mass, and thus on its density ρ_s, and consequently on the ratio X_4 (which reflects the influence of the specific mass ρ_s). On the other hand, if A is a property related to the totality of the moving grains (transport rate p, the size of sand waves, the effective roughness k_s of the mobile bed covered by sand waves, etc.), then the influence of the density ρ_s, and thus of X_4 on such properties, which are our main interest in practice, appears to be negligible. For example none of the existing transport formulae (with the exception of that of M. S. Yalin[7]) contains the ratio $X_4 = \rho_s/\rho$, other than that which contributes to the value of specific weight $\gamma_s = g\rho(\rho_s/\rho - 1)$, and which appears in the expression of

$$X_2 = \frac{\rho v_*^2}{\gamma_s D} = \frac{1}{\left(\dfrac{\rho_s}{\rho} - 1\right)} \frac{v_*^2}{gD}$$

This indicates clearly that the influence of X_4 on the transport rate must be barely noticeable.

The consideration of the specific weight γ_s alone (without specific mass ρ_s) seems to be sufficient to express the properties related to the totality of moving grains. Hence, if the investigation is confined to the properties related to the totality of the moving grains (to the transport of granular material en masse), then, with an accuracy sufficient for practical purposes, one can write

$$\Pi_A = \varphi_A(X_1, X_2, X_3) \tag{6.25}$$

One arrives then at the following reduced version of (6.9) (which no longer contains the condition $\lambda_{\rho_s} = \lambda_\rho$, that follows from $\lambda_{X_4} = 1$)

$$\left.\begin{aligned} \lambda_{v_*} &= \lambda_D^{-1} \\ \lambda_{v_*} &= \sqrt{(\lambda_D \lambda_{\gamma_s})} \\ \lambda_h &= \lambda_D \end{aligned}\right\} \tag{6.26}$$

However, even in this case, where we have only three conditions to satisfy five scales, we have not achieved much, so far as practical applications are concerned. Indeed, from the first and the third equations of (6.26) it follows that

$$\lambda_{v_*} = \frac{1}{\lambda_h} \tag{6.27}$$

On the other hand, considering that $v_* = \sqrt{(gSh)}$ we have

$$\lambda_{v_*} = \sqrt{(\lambda_S \lambda_h)} \tag{6.28}$$

and thus

$$\lambda_S = \frac{1}{\lambda_h{}^3} \tag{6.29}$$

(It is assumed, for the sake of simplicity, that the flow is uniform and thus that the energy gradient J coincides with the channel slope S.)

This equation (6.29) implies that even if the vertical model scale were as big as 1/5, it would be necessary to distort the slope 125 times! . . .

If the geometry of the granular material is specified, then two-dimensional transport of sediment is completely determined by the physical properties of the liquid and solid materials and by the mechanical structure of the two-dimensional flow. In equation form this may be written as

$$\text{Sediment transport} = f\,(\text{mechanics of flow, physical properties}) \tag{6.30}$$

Let us confine our considerations to the early stages of sediment transport where the granular material is moving in the vicinity of the bed (bed load). Clearly, in this case the transport of sediment will be determined by the mechanical structure of the flow in the vicinity of the bed, i.e. by the following distribution of u and τ_o

$$\left.\begin{aligned} u &= v_* \left(2{\cdot}5 \ln \frac{y}{k_s} + B_s\right) \\ \tau &\approx \tau_o = \rho v_*{}^2 \end{aligned}\right\} \tag{6.31}$$

$$\left(\text{where } B_s = \varphi_B \left(\frac{v_* k_s}{\nu}\right)\right)$$

But the expressions (6.31) indicate that the mechanical structure of the flow in the vicinity of the bed does not depend on h explicitly. Indeed, from (6.31) it follows clearly that

$$\text{mechanics of the flow} = f_1(v_*, \rho, \mu, k_s) \tag{6.32}$$

On the other hand, it is obvious that

$$\text{physical properties} = f_2(\rho, \mu, \gamma_s, D) \tag{6.33}$$

where the density ρ_s is deliberately not included, because we are dealing only with those properties of the phenomenon which are related to (or caused by) the transport of granular material en masse. Substituting (6.32) and (6.33) in (6.30) we obtain for the transport in the vicinity of the bed:

$$\text{transport of sediment} = f(f_1(v_*, \rho, \mu, k_s), f_2(\rho, \mu, \gamma_s, D))$$

i.e.

$$\text{transport of sediment} = \bar{f}(v_*, \rho, \mu, \gamma_s, D, k_s) \qquad (6.34)$$

which can be expressed in the dimensionless form as

$$\text{transport of sediment} = \varphi\left(X_1, X_2, \frac{k_s}{D}\right) \qquad (6.35)$$

where the form of the function φ, and the value of k_s/D are determined by the geometry of the surface of the mobile bed which, in general, is covered by sand waves. If the sand waves are not present (plain rough bed), then

$$\frac{k_s}{D} = \text{const}$$

and (6.35) reduces to

$$\text{transport of sediment} = \bar{\varphi}(X_1, X_2) \qquad (6.36)$$

The experiment shows that the size of sand waves is in general dependent on the flow depth h, and thus on the dimensionless variable X_3. (See Refs. 10 to 14.) Thus, even though the transport of sediment in the vicinity of the bed, as seen from (6.35) does not depend on X_3 directly it is dependent on X_3 indirectly by means of k_s/D.

Note that most of the existing transport formulae can be represented by the form

$$\Pi_p = \varphi_p(X_1, X_2)$$

which means that, strictly speaking, they can be valid only if the sediment is transported in the vicinity of a plain granular bed (regardless of whether their authors have explicitly mentioned this restriction or not).

From these explanations it follows that, since the simultaneous fulfilment of the requirements

$$\lambda_{X_1} = 1 \quad \lambda_{X_2} = 1 \quad \lambda_{X_3} = 1$$

(i.e. of the three conditions in (6.26)) is impossible, the only reasonable approach so far as transport in the vincinity of the bed is concerned, is to satisfy the conditions

$$\lambda_{X_1} = 1 \quad \lambda_{X_2} = 1 \qquad (6.37)$$

and that is precisely what is done (or is supposed to be done) nowadays, when carrying out model tests with granular material consisting of larger but lighter grains than the prototype. Indeed, the fulfilment of (6.37) implies the fulfilment of the first two conditions of (6.26), which can be expressed:

$$\left.\begin{aligned} \lambda_D &= \frac{1}{\lambda_{v_*}} \\ \lambda_{\gamma_s} &= \lambda_{v_*}{}^3 \end{aligned}\right\} \qquad (6.38)$$

and since λ_{v_*} implies $\quad \lambda_{v_*} = \sqrt{(\lambda_S \lambda_h)} = \sqrt{(n\lambda_y)}$

as

$$\left.\begin{aligned} \lambda_D &= \frac{1}{\sqrt{(n\lambda_y)}} \\ \lambda_{\gamma_s} &= \frac{1}{\lambda_D{}^3} \end{aligned}\right\} \qquad (6.39)$$

From (6.39) it follows that, since in all practical cases the product $n\lambda_y$ is smaller than unity, the scale λ_D is larger than unity while λ_{γ_s} is smaller than unity, or that the model bed material consists of larger but lighter grains than those of the prototype. This is a typical property of models with mobile bed, designed by taking into account the grain size Reynolds number X_1. Such models are widely used in contemporary laboratories if the study concerns a prototype having small-size bed material (or, to be more precise, if the order of the prototype Reynolds number $X_1' = v_*'D_{50}'/\nu$ is so small that the condition (6.17) cannot be satisfied). First systematic treatment of this kind of mobile bed model (having $\lambda_D > 1$; $\lambda_{\gamma_s} < 1$) is due to H. A. Einstein[15]. In spite of the apparent difference in the approach, the method of H. A. Einstein is essentially the same as that considered here[16]. In considering the similarity criteria of models it is not only wise, but necessary to use the dimensionless ratios

$$\frac{\bar{\gamma}_s}{\gamma} = \frac{\text{specific weight of material}}{\text{specific weight of water}}$$

and

$$\frac{\gamma_s}{\gamma} = \frac{\text{specific weight of material in water}}{\text{specific weight of water}}$$

By so doing we avoid specifying any particular system of units to define materials. The ratios are referred to as the relative specific weights of particular materials in air and water respectively. The relative specific weights $\bar{\gamma}_s/\gamma$ (in air) corresponding to some materials used as model sediment are given in Table 6.3.

Observe that the lightest model bed material used is polystyrene; its relative specific weight in water is $\gamma_s/\gamma = 0{\cdot}05$ to $0{\cdot}03$. In theory, there is, of course, no reason why an even lighter granular material should not be used. In practice, however, the use of a very light material is associated with certain difficulties. For example, it is difficult to make it sink, for a very small density difference often cannot overcome the surface tension produced by the oil film which usually surrounds the grain and which keeps it floating. In addition, a very light material is very sensitive to any

Table 6.3

Material	$\bar{\gamma}_s/\gamma$
Anthracite	1·40–1·70
Bakelite	1·35–2·05
Coal	1·20–1·50
Lignite	1·10–1·40
Nylon	1·14
Perspex	1·19
Polystyrene	1·03–1·05
P.V.C.	1·35–1·38
Silicon	2·40

occasional disturbance. For example, the model bed can be greatly disturbed when water first enters the model. Considering that the relative specific weight of sand or gravel (in water) is $\gamma_s/\gamma = 1{\cdot}65$, one arrives at the following range of practically possible values:

$$\frac{0{\cdot}03}{1{\cdot}65} = \frac{1}{55} \leqslant \lambda_{\gamma_s} < 1 \tag{6.40}$$

Using the inequality in the second equation of (6.38) we determine the following rather narrow range for λ_D

$$1 < \lambda_D < 3{\cdot}8 \tag{6.41}$$

It follows that the size of the model bed material cannot be more than 3·8 times larger than that of the prototype material. Considering that the present method should be used only if the prototype grain size is small (fractions of a millimeter), the fact that the model grains must be larger than the prototype grains hardly presents any difficulty in practice.

Let us now consider the following conditions (eqn. (5.61)), which must also be satisfied if the flow is non-uniform

$$\lambda_S = \frac{\lambda_y}{\lambda_x}, \quad \lambda_E = \frac{\lambda_y}{\lambda_x}, \quad \lambda_{\mathrm{Fr}} = 1 \tag{6.42}$$

As follows from 5.3.4, the value of E is the sum of the friction losses due to the effective roughness k_s of the flow boundaries, and of the local energy losses due to the variation in the geometry of the flow (sudden expansion, sudden contraction, bends, etc.). For the present, instead of considering the effective roughness k_s we will consider its following two components:

(i) the actual roughness of the surface of the mobile bed. Clearly, the value of this 'skin roughness' is of the order of the grain size D, regardless of whether the surface of the mobile bed is plain or undulated.

(ii) the roughness corresponding to the undulations, or irregularities, of the surface of mobile bed (i.e. that due to sand waves).

Let J_1, J_2 and J_3 be the components of the total energy gradient, $J = -\mathrm{d}H/\mathrm{d}x$, that are due to (i), (ii), and the variation in the geometry of the flow respectively. These three components can be expressed as follows:

$$\left.\begin{aligned} J_1 &= E_1 \,.\, \mathrm{Fr} \\ J_2 &= E_2 \,.\, \mathrm{Fr} \\ J_3 &= E_3 \,.\, \mathrm{Fr} \end{aligned}\right\} \tag{6.43}$$

where $(\mathrm{Fr} = v^2/gh)$

$$\left.\begin{aligned} E_1 &= \frac{1}{[2{\cdot}50 \times \ln(11X_3)]^2} \\ E_2 &\doteq \frac{1}{2} \times \frac{\Delta^2}{\Lambda h} \\ E_3 &= \zeta_L \times \frac{h}{L} \end{aligned}\right\} \tag{6.44}$$

For the sake of simplicity, E_1 is expressed by the simplest version of the friction factor corresponding to the rough turbulent flow. However, one can always substitute the multiplier 11 by a certain function of X_1, and express all the relations that follow accordingly. The derivation of the expression for E_2 can be found elsewhere[8,17].

Here Δ and Λ are the height and length of sand waves; the meaning of ζ_L being clear from (5.53). Summing the equations (6.44) we obtain

$$J = J_1 + J_2 + J_3 = (E_1 + E_2 + E_3)\,\mathrm{Fr} \tag{6.45}$$

where the expression in the brackets is the total energy loss factor E. Hence, the second condition of (6.42) can be expressed as follows:

$$\lambda_E = \frac{E_1'' + E_2'' + E_3''}{E_1' + E_2' + E_3'} = \frac{\lambda_y}{\lambda_x} \tag{6.46}$$

As has been explained in 5.3.4 and indicated by the relations (5.57) to (5.58) the equation

Eq. (5.61)

$$E_3'' = \frac{\lambda_y}{\lambda_x} E_3' \tag{6.47}$$

can be regarded as being satisfied automatically. Accordingly, the fulfilment of the condition (6.46) implies the fulfilment of

$$(E_1'' + E_2'') = \frac{\lambda_y}{\lambda_x} (E_1' + E_2') \tag{6.48}$$

If the bed is plain, i.e. if $\Delta = 0$, then (as is obvious from (6.44)) $E_2 = 0$, while (6.48) reduces into

$$E_1'' = \frac{\lambda_y}{\lambda_x} E_1' \tag{6.49}$$

which can be expressed as

$$\frac{\ln (11X_3')}{\ln (11X_3'')} = \sqrt{\left(\frac{\lambda_y}{\lambda_x}\right)}$$

or as

$$\sqrt{\frac{\lambda_y}{\lambda_x}} \left(1 - \frac{\ln\left(\dfrac{1}{\lambda_{X_3}}\right)}{\ln (11X_3')}\right) = 1 \tag{6.50}$$

Observe that the value of λ_{X_3} depends on the prototype value $X_3' = h'/D_{50}'$. In the following considerations, it will be assumed that (6.49) is satisfied for those stages when the surface of the mobile bed is plain. In other words, it will be assumed that X_3' is formed by an h' which is near to h_{cr}', corresponding to the initiation of sediment transport on the plain bed (i.e. corresponding to $E_2 = 0$). But, if for the case of a plain bed (6.50) is valid, and thus (6·49) is fulfilled, then the condition (6.46) can also be regarded as fulfilled, provided that the sand waves are either not present or they have a negligible height (in comparison to their length and flow depth). If, on the other hand, the sand waves are developed, i.e. if the value of

$$E_2 = \frac{1}{2} \frac{\Delta^2}{\Lambda h}$$

is appreciable (in comparison to $E_1 + E_3$), then the fulfilment of the condition (6.46) will, in addition, depend on the fulfilment of

$$E_2'' = \frac{\lambda_y}{\lambda_x} E_2' \tag{6.51}$$

i.e. of

$$\frac{\Delta''^2}{\Lambda'' h''} = \frac{\lambda_y}{\lambda_x} \cdot \frac{\Delta'^2}{\Lambda' h'} \tag{6.52}$$

In the following we will assume that the flow is tranquil (Fr < 1) and thus that the sand waves are either *ripples* or *dunes*, and introduce the following rule of thumb: if $X_1 < \approx 20$, then the sand waves are ripples; if $X_1 > \approx 20$ then, they are dunes.

(i) If the sand waves are ripples (i.e. if $v_* D_{50}/\nu < \approx 20$), then their size can be regarded as proportional to the grain size D (and independent of the flow depth h). In practice, the conditions are not so simple and, strictly speaking, this rule is valid only if $X_3 = h/D$ is sufficiently small (smaller than ≈ 1000 say). If X_3 is large, then sand waves occur (in a tranquil flow) as follows:

only ripples if $X_1 < \approx 8$

only dunes if $X_1 > \approx 24$, and

ripples superimposed on dunes if $\approx 8 < X_1 < \approx 24$

(For more extensive information on this subject see Ref. 18.)

$$\Delta \sim D \quad \text{and} \quad \Lambda \sim D \tag{6.53}$$

The proportionality factors in the relations (6.53) (which correspond to ripples) can only be the functions of X_1 and X_2. Since, in the case of the model under consideration, the model and prototype values of X_1 and X_2 are identical, the proportionality factors in (6.53) are also identical. This holds for any stage of the flow (for the conditions $\lambda_{X_1} = 1$ and $\lambda_{X_1} = 1$ are fulfilled independently of the stage).

From (6.53) it follows that

$$\lambda_\Delta = \lambda_\Lambda = \lambda_D \tag{6.54}$$

and consequently that the model and prototype values of $E_2 = \Delta^2/\Lambda h$ are related to each other as

$$\frac{\Delta''^2}{\Lambda'' h''} = \frac{\lambda_\Delta^2}{\lambda_\Lambda \lambda_y} \cdot \frac{\Delta'^2}{\Lambda' h'} = \frac{\lambda_D}{\lambda_y} \frac{\Delta'^2}{\Lambda' h'}$$

i.e. as

$$E_2'' = \frac{\lambda_D}{\lambda_y} E_2' \tag{6.55}$$

Hence, the fulfilment of (6.51) is equivalent to that of $\lambda_y/\lambda_x = \lambda_D/\lambda_y$, i.e. of

$$\lambda_D = \frac{\lambda_y{}^2}{\lambda_x} \tag{6.56}$$

Equating this value of λ_D with the value

$$\lambda_D = \frac{1}{\sqrt{(n\lambda_y)}}$$

given by the first eqn. (6.39) we arrive at

$$n\lambda_y = 1 \tag{6.57}$$

Observe that the eqn. (6.57) implies that if the vertical model scale is as reasonable as, say, 1/20, then the distortion must be as unreasonable as $n = 20$. It follows that if the sand waves are ripples, the condition (6.51) cannot be satisfied in practice.

(ii) If the sand waves are *dunes* (i.e. if $v_* D_{50}/\nu > \approx 20$), then their size can be regarded as proportional to the flow depth h (and independent of the grain size D) as

$$\Delta \sim h \quad \text{and} \quad \Lambda \sim h \tag{6.58}$$

From (6.58) it follows that

$$\lambda_\Delta = \lambda_\Lambda = \lambda_y \tag{6.59}$$

and thus that the model and prototype values of $E_2 = \Delta^2/\Lambda h$ are related to each other as

$$\frac{\Delta''^2}{\Lambda'' h''} = \frac{\lambda_\Delta{}^2}{\lambda_\Lambda \lambda_y} \cdot \frac{\Delta'^2}{\Lambda' h'} = \frac{\Delta'^2}{\Lambda' h'}$$

i.e. as

$$E_2'' = E_2' \tag{6.60}$$

Hence, the requirement of (6.51) is equivalent to that of

$$\frac{\lambda_y}{\lambda_x} = 1 \tag{6.61}$$

which is incompatible with the fact that the model under consideration is a distorted one.*

* The proportionality factors in the relations (6.58) (which correspond to dunes) are, in general, the functions of X_1, X_2, and X_3. They become independent of X_3 only if at least one of the variables X_1 and X_3 is sufficiently large (if $X_1 > \approx 40$ and/or $X_3 > \approx 1000$). Since the value of X_3 is not the same in model and prototype, the treatment above (i.e. the steps from (6.58) to (6.61)), strictly speaking, can be regarded as correct only if X_1 and/or X_3 are sufficiently large.

The result (6.61) does not, however, mean that the dunes cannot be reproduced in a distorted model; all it means is that the energy loss due to dunes cannot be reproduced in a required manner by (6.51). In Example 6.4, it will be shown that the dunes themselves can be reproduced in a distorted model in a perfectly satisfactory manner.

From these considerations, we arrive at the conclusion that the condition (6.51), cannot in general, be satisfied by the distorted model under consideration (having larger but lighter bed material). This brings us to the question: "What will actually happen if (6.46) is not satisfied?" In order to answer this, let us consider the relation (6.46) which implies that each E_j'' must be n times larger than its prototype counterpart E_j'. Since (6.47) can be regarded as valid, and since λ_{X_3} is given by (6.50) (which follows from (6.46)), the requirements

$$E_3'' = nE_3' \quad \text{and} \quad E_1'' = nE_1'$$

can be regarded as satisfied. The fulfilment of these two requirements is perfectly sufficient to satisfy the condition (6.46) before and at the beginning of, sediment transport, that is, when the bed is (or can be regarded as) plain ($E_2 = 0$). With the beginning of sediment transport, the development of sand waves begins, and consequently the increment of the coefficient E_2 (from zero onwards). However, as is clear from inequalities below, that λ_D/λ_y is larger than n,

$$\left.\begin{aligned} &E_2'' = \frac{\lambda_D}{\lambda_y} E_2' > nE_2' &&\text{(for ripples)} \\ &\text{and since} \\ &E_2'' = E_2' < nE_2' &&\text{(for dunes)} \end{aligned}\right\}^{*} \qquad (6.62)$$

(which follows from (6.48), (6.52) and (6.57)) the term E_2'' does not grow at the required rate, in the case of both ripples and dunes. Consider, for example, the case of dunes.

With the development of sand waves, the model will appear to be smoother than it should be. As the result of this smoothness, the free surface of the model flow will tend to remain below the position required. This situation is shown schematically in Fig. 6.4. When the bed is plain, then the condition (6.46) is satisfied; the energy loss is dynamically similar, and therefore, the model free surface is in the required position (Fig. 6.4a). When the dunes are developed, then E_2'' is less than nE_2', and thus E'' is less than nE'; the energy loss per unit length of the model is less than necessary, and consequently the free surface of the (tranquil) flow deviates progressively from the required position when x increases (Fig. 6.4b). In

* It is assumed, in accordance with reality, that λ_D/λ_y is larger than n.

practice this deviation is eliminated, or at least effectively reduced, by adjusting the tail gate at the downstream end of the model.

From experimental measurements carried out in the field and the laboratory, it is established that the *steepness* Δ/Λ of the sand waves does not usually exceed $\approx 1/10$, while their relative height Δ/h is seldom more than $\approx 1/5$. Hence, with a reasonable accuracy for practical purposes, one can write

$$(E_2)_{\max} \approx \frac{1}{2} \times \frac{1}{10} \times \frac{1}{5} = \frac{1}{100}$$

On the other hand, as is well-known from hydraulics, the values of ζ_i (and thus of E_3) corresponding to ordinary local energy losses can exceed

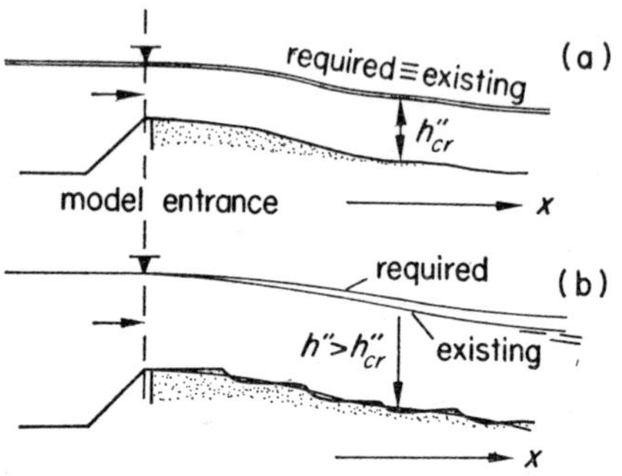

FIG. 6.4

many times the maximum value of E_2 shown above. Thus, the error due to the fact that E_2'' is less than the required value nE_2' will certainly lose importance as the irregularities in the general pattern of the flow increase (meandering, intensive variation of the flow cross-section, etc.).

The determination of scales for a distorted model with a mobile bed considered in the present section can be summarized as follows: We have five basic conditions

$$\lambda_{X_1} = 1; \quad \lambda_{X_2} = 1; \quad \lambda_S = n; \quad \lambda_E = n; \quad \lambda_{\mathrm{Fr}} = 1$$

which result in the following five equations

$$\left.\begin{aligned}
&\lambda_{X_1} = 1 \to \lambda_D = \frac{1}{\sqrt{(\lambda_y n)}} \\
&\lambda_E = n \to \sqrt{n}\left(1 - \frac{\ln(\lambda_D/\lambda_y)}{\ln(11X_3')}\right) = 1 \\
&\lambda_{X_2} = 1 \to \lambda_{\gamma_S} = (\lambda_y n)^{\frac{3}{2}} \\
&\lambda_S = n \to \lambda_S = n \\
&\lambda_{\mathrm{Fr}} = 1 \to \lambda_v = \sqrt{(\lambda_y)}
\end{aligned}\right\} \qquad (6.63)$$

If, from the first two equations the values of λ_D, λ_y and n are determined, then the values of $\lambda_{\gamma s}$, λ_S and λ_v follow immediately from the last three equations. In other words, the problem virtually is solved if *three* scales (λ_D, λ_y and n) are determined from the first *two* equations. As is clear from the second equation (and as has been mentioned before), the value of the determined scales will be dependent on the selected prototype value for $X_3' = h'/D_{50}'$. In the preceding considerations, it was assumed that X_3' is formed by the critical depth $h' = h_{cr}'$. However, this does not necessarily need to be so. If, for example, the investigation is restricted to a particular stage only (or to a particular range of stages), then there is no reason why X_3' should not be formed by the flow depth corresponding to that stage (or representing that range of stages). On the other hand, if there is no preference for a particular range of stages, and yet if the knowledge of when and where the transport of sediment begins is important, then it would certainly be advisable to have the most accurate reproduction of the phenomenon in the vicinity of its critical stage, and thus to have X_3' formed by $h' = h_{cr}'$.

It should be pointed out that even though the solution of the problem does vary, depending on the choice of X_3', fortunately this variation is rather weak, and therefore, the solution obtained for a particular value of X_3', i.e. of h' can usually be regarded as valid for a considerable range of flow depths. The reason for this lies in the fact that the values of X_3' are usually very large (several thousands), and the logarithmic function is insensitive to variations in large values of its variable. For example,

$$\text{if } X_3' = 5000 \quad \text{then} \quad \ln(11 \times 5000) = 10{\cdot}9$$
$$\text{if } X_3' = 10\,000 \quad \text{then} \quad \ln(11 \times 10\,000) = 11{\cdot}6$$

Hence, the variation of the flow depth by a factor of 2 induces a variation in $\ln(11X_3')$ of only 6·5%, which shows that the importance of the correct selection of X_3' should not be exaggerated. The same does not apply to $\ln(\lambda_D/\lambda_y)$, for λ_D/λ_y is of the order ≈ 10 (not of thousands).

Let us now see the application of eqns. (6.63) on some examples.

EXAMPLE 6.3

Consider the flow in a river which has an average cross-section shown in Fig. 6.5 and which possesses the following characteristics

$$Q_{max}' = 254{\cdot}5 \text{ m}^3/\text{s},\ \omega_{max}' = 702 \text{ m}^2,\ h_{max}' = 5{\cdot}20 \text{ m},\ B_{max} = 150 \text{ m}$$
$$Q_{min}' = 38{\cdot}5 \text{ m}^3/\text{s},\ \omega_{min}' = 213 \text{ m}^2,\ h_{min}' = 1{\cdot}70 \text{ m}$$
$$S' = 0{\cdot}00001,\ D_{50}' = 0{\cdot}30 \text{ mm}$$
$$(g = 9{\cdot}81 \text{ m/s}^2;\ \nu = 10^{-6} \text{ m}^2/\text{s};\ \gamma_s'/\gamma = 1{\cdot}65)$$

Design the model with a mobile bed.

The width/depth ratio is sufficiently large even when $Q' = Q_{max}'$ ($150/5{\cdot}20 = 28{\cdot}8$). Hence, in the central region of the cross-section, the flow can be treated as a two-dimensional flow. Let us assume that it is intended to form X_3' by $h' = h_{cr}'$. Thus, the order of h_{cr} must be determined. From the Shields curve, in Fig. 6.2, we find for sand (of $D_{50}' = 0{\cdot}30$ mm)

$X_3 = \frac{h}{D}$

$$(X_1)_{cr} = 4{\cdot}25$$
$$(X_2)_{cr} = 0{\cdot}04$$

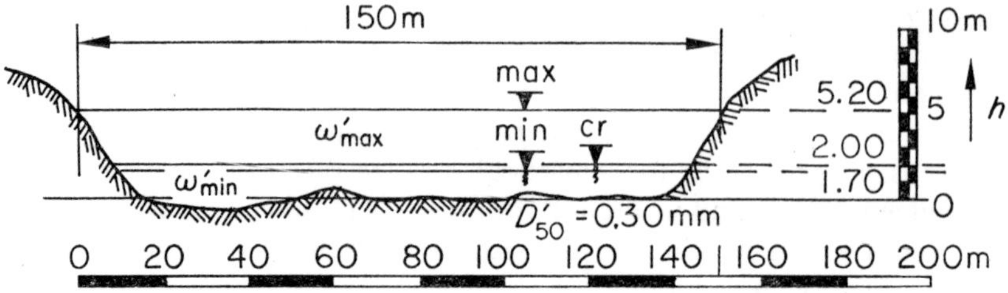

FIG. 6.5

Using any one of these values, one can determine h_{cr}'. For example, using $(X_1')_{cr} = 4{\cdot}25$, using the slope S' as the average value of the energy gradient J', on the grounds mentioned in the text following the eqns. (5.18) and (5.19), we obtain

$$\frac{\sqrt{(gS'h_{cr}')}D_{50}'}{\nu} = \frac{\sqrt{(9{\cdot}81 \times h_{cr}' \times 10^{-5})}0{\cdot}3 \times 10^{-3}}{10^{-6}} = 4{\cdot}25$$

and thus

$$h_{cr}' = 2{\cdot}00 \text{ m}$$

Hence, the bed material will be transported at all flow depths greater than 2·00 m. Observe that

$$(X_1')_{max} = 4{\cdot}25 \sqrt{\left(\frac{5{\cdot}20}{2{\cdot}00}\right)} = 6{\cdot}85$$

and thus that the order of X_1' is smaller than ≈ 20 throughout the stages of the flow, which implies that the sand waves covering the bed of the flow are always ripples. Using $h_{cr}' = 2{\cdot}00$ m we determine the following values for X_3'

$$X_3' = \frac{2000}{0{\cdot}3} = 6670$$

Thus

$$\ln(11X_3') = \ln(11 \times 6670) = 11{\cdot}2$$

and consequently the second equation in (6.63) can be expressed as follows:

$$\sqrt{n}\left(1 - \frac{\ln(\lambda_D/\lambda_y)}{11\cdot 2}\right) = 1$$

This equation together with

$$\lambda_D = \frac{1}{\sqrt{(\lambda_y n)}}$$

which must satisfy $1 < \lambda_D < 3\cdot 8$ (see (6·41)) form the necessary system in order to determine any two of λ_y, λ_D and n if one of these three quantities is chosen freely. It seems reasonable to have in the model, say

$$h_{cr}'' = 10 \text{ cm and } h_{max}'' = \frac{5\cdot 20}{2\cdot 00} \times 10 = 26\cdot 0 \text{ cm}$$

which means that it would be desirable to have

$$\lambda_y = \frac{10}{200} = \frac{1}{20} = 0\cdot 05$$

Substituting this value in the equations above, and eliminating λ_D, we arrive at the following equation determining the distortion n

$$\sqrt{n}\left(1 - \frac{\ln(20^3/n)}{22\cdot 4}\right) = 1$$

The equation is transcendent and thus the value of n cannot be solved directly; using the trial and error method we determine

$$n = 2\cdot 5$$

Using this value of the distortion, together with $\lambda_y = 1/20$, in the expression for λ_D we obtain

$$\lambda_D = \sqrt{\left(\frac{20}{2\cdot 5}\right)} = \sqrt{8} = 2\cdot 83$$

which satisfies $1 < \lambda_D < 3\cdot 8$. Otherwise we should choose a different value for λ_y if we do not wish to have in the model a bed material lighter than polystyrene.

The substitution of

$$\lambda_y = \frac{1}{20}, \quad n = 2\cdot 5 \quad \text{and} \quad \lambda_D = 2\cdot 83$$

in (6.63) gives the following values for the remaining scales

$$\lambda_{\gamma_s} = \frac{1}{2{\cdot}83^3} = \frac{1}{22{\cdot}6} = 0{\cdot}0443$$

$$\lambda_S = 2{\cdot}5$$

$$\lambda_v = \frac{1}{\sqrt{20}} = \frac{1}{4{\cdot}47} = 0{\cdot}224$$

Hence, the (geometrically similar) model bed material is given by

$$D_{50}'' = 2{\cdot}83 \times 0{\cdot}3 = 0{\cdot}85 \text{ mm}$$

and

$$\gamma_s''/\gamma = 1{\cdot}65 \sqrt{0{\cdot}0443} = 0{\cdot}073$$

which implies that the relative specific weight of the model granular material (in air) must be

$$\bar{\gamma}_s''/\gamma = 1{\cdot}073$$

If the required value of $\bar{\gamma}_s''$ does not coincide with any of specific weights of the materials available in commerce, then one must choose the nearest available and adjust the values of λ_y and of the rest of the scales accordingly.

The prototype average velocities are

$$v_{\max}' = 702/254{\cdot}5 = 0{\cdot}36 \text{ m/s}$$

and

$$v_{\min}' = 38{\cdot}5/213 = 0{\cdot}18 \text{ m/s}$$

Using these values, together with the values of prototype characteristics given at the beginning of the example, we arrive at the following set of essential characteristics of the model flow

$$h_{\max}'' = 0{\cdot}05 \times 5{\cdot}20 = 0{\cdot}26 \text{ m} \quad (26{\cdot}0 \text{ cm})$$
$$h_{\min}'' = 0{\cdot}05 \times 1{\cdot}70 = 0{\cdot}085 \text{ m} \quad (8{\cdot}5 \text{ cm})$$
$$Q_{\max}'' = 0{\cdot}224 \times 0{\cdot}001 \times 254{\cdot}5 = 56{\cdot}75 \times 10^{-3} \text{ m}^3/\text{s} \quad (56{\cdot}75 \text{ l/s})$$
$$Q_{\min}'' = 0{\cdot}224 \times 0{\cdot}001 \times 38{\cdot}5 = 8{\cdot}62 \times 10^{-3} \text{ m}^3/\text{s} \quad (8{\cdot}67 \text{ l/s})$$
$$v_{\max}'' = 0{\cdot}224 \times 0{\cdot}36 = 0{\cdot}081 \text{ m/s}$$
$$v_{\min}'' = 0{\cdot}224 \times 0{\cdot}18 = 0{\cdot}04 \text{ m/s}$$
$$S'' = 2{\cdot}5 \times 10^{-5}$$
$$B_{\max}'' = 0{\cdot}02 \times 150 = 3{\cdot}00 \text{ m}$$

The multiplier 0·02 in the expression of $B_{\max}''$ is the horizontal scale

$$\lambda_x = \frac{\lambda_y}{n} = \frac{1}{20 \times 2{\cdot}5} = \frac{1}{50}$$

The multiplier 0·001 in the expressions of Q_{max}'' and Q_{min}'' being

$$\lambda_\omega = \lambda_x \lambda_y = \frac{1}{50}\frac{1}{20}$$

The flow in the river is always turbulent. Accordingly, it must be so in the model (even when it is operating with its smallest depth and velocity). In other words, one must check that the vertical model scale is not too small. The minimum value of the Reynolds number of the model flow is given by

$$\mathrm{Re}_{min}'' = \frac{v_{min}'' \times h_{min}''}{\nu}$$

Substituting the corresponding values we obtain

$$\mathrm{Re}_{min}'' = \frac{0{\cdot}04 \times 0{\cdot}085}{10^{-6}} = 3440$$

which is considerably higher than the order 500 to 1000 corresponding to the conversion from laminar to turbulent in open channels. The cross-sections of all circular pipes are geometrically similar; the channel cross-sections are not. Hence, the existence of a much wider range for the values of the critical Reynolds number (corresponding to the conversion). The fact that the order of the critical Reynolds number for channel flows is less than $\approx$2300 (corresponding to the flow in a circular pipe) can be attributed to the presence of the free surface, which is considerably less effective than the rigid boundary in damping out the disturbances.

EXAMPLE 6.4

In the previous example the sand waves were ripples. Let us now consider the case of a mobile bed covered by dunes. The Table 6.4 contains the characteristics of dunes in the Mississippi river (having more than 10 metres flow depth) together with the characteristics of dunes formed in a laboratory flume, Plate 4 (having a depth of a few centimetres). In spite of the fact that the flow depths are more than 200 times different, and that in one flow the bed material is sand, whereas in the other it is polystyrene, the geometry presented by the dunes is remarkably similar. Indeed, the dimensionless aspects Λ/h, Δ/Λ and Δ/h, characterizing the dune geometry are of the same order for both flows. The fact that this similarity is present, in spite of the fact that the slope of the laboratory flow is more than twice as steep as that of the prototype, is an additional encouragement, for the slope of a distorted model, has to be n times steeper than the prototype slope.

Let us assume that we want to reproduce this region of the Mississippi river in a model with a mobile bed. We choose as the bed material polystyrene ($\bar{\gamma}_s''/\gamma = 1{\cdot}03$), that is, we select the specific weight scale as

$$\lambda_{\gamma_s} = \frac{0{\cdot}03}{2{\cdot}65} = \frac{1}{55}$$

which gives

$$\lambda_D = \sqrt[3]{55} = 3{\cdot}8$$

Table 6.4

	$S \times 10^5$	h (m)	Λ (m)	Δ (m)	$\frac{\Lambda}{h}$	$\frac{\Delta}{\Lambda}$	$\frac{\Delta}{h}$
Mississippi river[19] $D_{50} = 0{\cdot}27$ to 0·56 mm (sand)	≈10	25·20	122·00	2·74	4·85	0·0225	0·109
	≈10	17·70	122·00	3·66	6·80	0·0297	0·209
		(cm)	(cm)	(cm)			
2 feet wide laboratory flume $D = 1{\cdot}35$ mm* (polystyrene)	23·33	9·26	67·0	1·31	7·24	0·0194	0·140
	23·33	8·12	42·4	1·24	5·21	0·0292	0·151

* Uniform material.

Accordingly, the model grain size $D_{50}'' = 1{\cdot}35$ mm corresponds to the prototype grain size

$$D_{50}' = \frac{1{\cdot}35}{3{\cdot}8} = 0{\cdot}355 \text{ mm}$$

This value of the prototype grain size can perfectly well represent the size of the Mississippi sand having the diameter D_{50}' varying between 0·27 and 0·56 mm.

Let us suppose that the typical value of the flow depth is

$$h' = 15{\cdot}0 \text{ m } (\approx 49 \text{ ft})$$

In this case, the typical value of X_3' must be

$$X_3' = \frac{15\,000}{0{\cdot}355} = 42\,250$$

while

$$\ln (11 \times X_3') = 13{\cdot}04$$

Hence, the values of λ_y and n have to be determined by solving

$$\sqrt{(n\lambda_y)} = \frac{1}{3{\cdot}8}$$

and

$$\sqrt{n}\left(1 - \frac{\ln (3{\cdot}8/\lambda_y)}{13{\cdot}04}\right) = 1$$

Eliminating $\sqrt{n}$ we arrive at the following equation for λ_y alone

$$\frac{1}{\sqrt{\lambda_y}}\left(1 - \frac{\ln (3{\cdot}8/\lambda_y)}{13{\cdot}04}\right) = 3{\cdot}8$$

which gives (by trial and error)

$$\lambda_y = \frac{1}{37{\cdot}5}$$

and consequently

$$n = \frac{37{\cdot}5}{(3{\cdot}8)^2} = 2{\cdot}60; \quad \left(\lambda_x = \frac{1}{37{\cdot}5 \times 2{\cdot}60} = \frac{1}{97{\cdot}5}\right)$$

The rest of the scales in (6.63) are

$$\lambda_S = 2{\cdot}60$$

and

$$\lambda_v = \frac{1}{\sqrt{37{\cdot}5}} = \frac{1^*}{6{\cdot}125}$$

Accordingly, the model flow depth (corresponding to the prototype value, say $h' = 20$ m) will be

$$h'' \approx \frac{2000}{37{\cdot}5} = 53{\cdot}3 \text{ cm}$$

while the model width (corresponding to the prototype order, say $B = 1$ km)

$$B'' \approx \frac{1000}{97{\cdot}5} = 10{\cdot}25 \text{ m}$$

* Observe that $\lambda_s = 2{\cdot}60$ gives, for the model slope, $S'' = 26{\cdot}10^{-5}$, which is very near the laboratory flume slope $23{\cdot}33 \times 10^{-5}$ in Table 6.4.

This is a rather big model, but not impossibly so. Models of this size are usually built outdoors. Suppose the length of the river region which must undergo the model tests is $L' = 15$ km. In this case, the length of the (outdoor) model would be

$$L'' = \frac{15\,000}{97{\cdot}5} = 154 \text{ m}$$

The expenses of a model (however big) are usually generously compensated for by the information received, and consequently by the achievement of the most effective and economic designs.

6.6 Scales of the Bed Material Motion

Before closing this chapter, attention should be drawn to the following relevant properties of the distorted model studied here.

6.6.1 Length and Time Scales of Individual Grain Motion The length which appears in the dimensionless variables X_1 and X_2 which determine the progress of the two-phase phenomenon, in the vicinity of the bed, is the grain size D. Thus the geometric (linear) scale of dynamic similarity in the vicinity of the bed (given by $\lambda_{X_1} = 1$ and $\lambda_{X_2} = 1$) is λ_D (not λ_y or

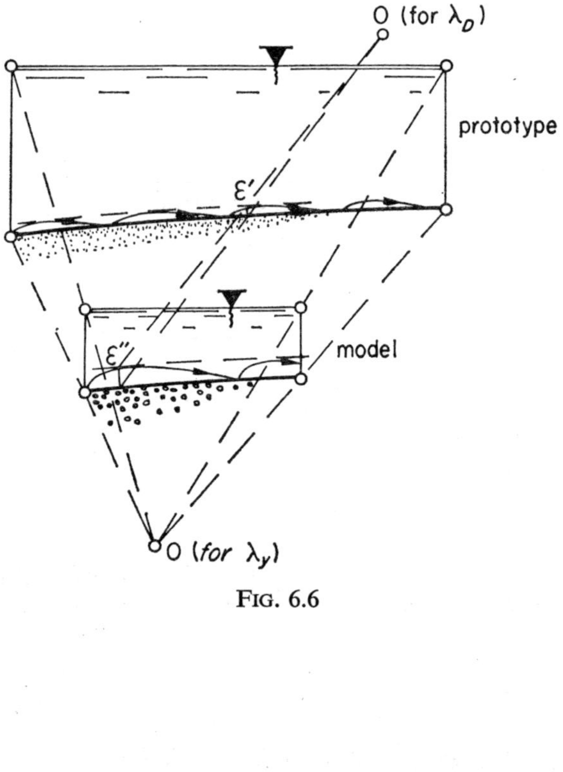

FIG. 6.6

7 Napier Harbour model, Hydraulic Research Station, Wallingford

8 Tauranga Harbour model, Hydraulics Research Station, Wallingford

λ_x). This implies a certain distortion in the behaviour of the grains (if the model is analysed according to the model scale λ_y). Indeed, the *linear* properties related to the motion of individual grains are not λ_y times *smaller*; they are λ_D times *larger*. For example, the height and the length of the trajectory of a saltating grain in the model is λ_D times larger than in the prototype (Fig. 6.6). Thus, the thickness ε occupied by the moving grains is also λ_D times larger. Similarly, there is a difference in the time scale. Consider, for example, the time scale of the fluid motion

$$\lambda_t = \frac{\lambda_x}{\lambda_{U_x}} = \frac{\lambda_x}{\sqrt{\lambda_y}} = \frac{\sqrt{\lambda_y}}{n}$$

This scale is not equal to the time scale $(\lambda_t)_s$ of the grain motion, for the velocity which appears in the dimensionless variables X_1 and X_2 of the grain motion is v_*, while the length is D. Thus, the typical time of the grain motion is

$$\frac{D}{v_*}$$

while its scale $(\lambda_t)_s$ is

$$(\lambda_t)_s = \frac{\lambda_D}{\lambda_{v_*}} = \frac{1}{\lambda_{v_*}^{\ 2}} = \frac{1}{n\lambda_y} \tag{6.64}$$

Observe that

$$\frac{(\lambda_t)_s}{\lambda_t} = \frac{1}{\lambda_y^{\frac{3}{2}}} \gg 1$$

and thus

$$(\lambda_t)_s \gg \lambda_t$$

is valid. Hence, the motion of the model grains is delayed compared with the motion of the fluid.

6.6.2 Scales of the Properties of the Grain Motion en masse When determining the scale λ_A of a property A related to the motion of grains en masse (in the vicinity of the bed), it would be advisable first to determine λ_A in terms of the scales of the characteristic parameters describing the two phase motion in the vicinity of the bed, and subsequently to express the scales of the characteristic parameters (forming λ_A) in terms of the model scales. Let us consider, for example, the scale of the transport rate P

(the weight of granular material passing per unit time through the total width b of the channel bed). Clearly

$$P = bp$$

where p is the specific transport rate (per unit width of the flow). Thus

$$\lambda_P = \lambda_b \lambda_p = \lambda_x \lambda_p \tag{6.65}$$

Now p, being the property of the grain motion in the vicinity of the bed, its scale must be expressed by means of the scales of the characteristic parameters

$$\mu, \rho, D, v_*, \gamma_s$$

Accordingly, we can have

$$\left.\begin{aligned} &\lambda_p = \lambda_{\gamma_s} \lambda_D \lambda_{v_*} \\ \text{or} \quad &\lambda_p = \lambda_\rho \lambda_{v_*}{}^3 \\ \text{or} \quad &\lambda_p = \lambda_\rho{}^{\frac{1}{2}} \lambda_{\gamma_s}{}^{-\frac{3}{2}} \lambda_D{}^{-\frac{3}{2}} \end{aligned}\right\} \tag{6.66}$$

The reader can easily prove that, whichever version is selected, by substituting for the scales of characteristic parameters on the right hand sides, their values in terms of λ_y and n, the same value is obtained.

$$\lambda_p = (n\lambda_y)^{\frac{3}{2}} \tag{6.67}$$

Hence

$$\lambda_P = \lambda_x (n\lambda_y)^{\frac{3}{2}} = \lambda_y{}^{\frac{5}{2}} n^{\frac{1}{2}} \tag{6.68}$$

Consider another example. Suppose that from a small area $\bar{\omega}''$ of the surface of the model bed, N_D'', grains are detaching during the time t''. How many grains are detaching in the prototype from the corresponding area per unit time? Since the number N_D of detachments must be proportional to the total number of the grains in the area $\bar{\omega}$ and to the time t, we have

$$N_D \sim \frac{\bar{\omega}}{D^2} \cdot t \tag{6.69}$$

and thus

$$\lambda_{N_D} = \frac{\lambda_\omega}{\lambda_D{}^2} (\lambda_t)_s \tag{6.70}$$

Substituting here

$$\lambda_{\bar{\omega}} = \lambda_x \lambda_y, \quad \lambda_D{}^2 = \frac{1}{n\lambda_y}, \quad (\lambda_t)_s = \frac{1}{n\lambda_y}$$

we obtain

$$\lambda_{N_D} = \lambda_x \lambda_y \tag{6.71}$$

i.e.

$$N_D' = \frac{N_D''}{\lambda_x \lambda_y} \tag{6.72}$$

This is the number of detachments during the time $t' = t'' n \lambda_y$; the number $(N_D')_1$ of detachments per unit time being

$$(N_D')_1 = \frac{N_D'}{t'} = \frac{N_D''}{t''} \frac{1}{y_y{}^2 \lambda_x n}$$

i.e.

$$(N_D')_1 = \frac{1}{\lambda_y{}^3} \frac{N_D''}{t''} \tag{6.73}$$

(Observe that if $\lambda_y = 1/30$, then $(N_D')_1$ is $30^3 = 27\,000$ times larger than $N_D''/t'' = (N_D'')_1$.)

6.6.3 Time Scales Related to the Formation of the Mobile Bed Surface The following considerations are carried out for the directions y and x perpendicular and parallel to the flow respectively.

(i) *Formation in the direction of y-axis.* Let Y be the elevation of the surface of the bed corresponding to a certain location (x) and instant (t). The variation of Y with the time t is related to the variation of transport rate p with the distance x by the following differential equation (of Polia)

$$-\frac{\partial Y}{\partial t} = \frac{1}{\gamma_s} \frac{\partial p}{\partial x} \tag{6.74}$$

which gives immediately the scale relation

$$\lambda_Y = \frac{\lambda_t \lambda_p}{\lambda_{\gamma_s} \lambda_x} \tag{6.75}$$

where λ_t must obviously be identified with $(\lambda_t)_s$ given by (6.64). Using for λ_t the value (6.64) while for λ_p, the first relation in (6.66) we obtain

$$\lambda_Y = \frac{\lambda_D{}^2}{\lambda_x} \tag{6.76}$$

This is the scale of displacement of the surface of the bed in the vertical direction for the time intervals related to each other by the scale $(\lambda_t)_s$.

What is the time scale $(\lambda_t)_y$ that would yield the model displacement Y'' in vertical model scale?

Clearly, the answer to this question is given by the following relation

$$\lambda_y = \frac{(\lambda_t)_y}{(\lambda_t)_s} \lambda_Y \tag{6.77}$$

Substituting here the values $(\lambda_t)_s$ and λ_Y given by (6.64) and (6.76), and taking into account that

$$\lambda_D = \frac{1}{\lambda_{v_*}}$$

we arrive at

$$(\lambda_t)_y = \lambda_y \lambda_x = \frac{\lambda_y{}^2}{n} \tag{6.78}$$

This is the time scale that should be used when predicting the changes in the elevation of the surface of the bed (erosion-accretion).

(ii) *Formation in the direction of x-axis.* What is the time scale $(\lambda_t)_x$ corresponding to the transport of granular material in the direction x of the flow? Let us assume that during the prototype time $\delta t'$ the prototype material moves from a location x' to a location $x' + \delta x'$. During the model time $\delta t''$ the model material travels the distance $\delta x''$. Clearly, if $\delta t''$ and $\delta t'$ are related by

$$\frac{\delta t''}{\delta t'} = (\lambda_t)_s \tag{6.79}$$

then $\delta x''$ and $\delta x'$ must be related by

$$\frac{\delta x''}{\delta x'} = \lambda_D \tag{6.80}$$

for the linear scale of the (saltating or rolling) grain motion, corresponding to $(\lambda_t)_s$ is λ_D. If, on the other hand, it is desired to have the displacement scale equal to λ_x, i.e. to be λ_x/λ_D times different than the value given by (6.80), then the ratio of the durations must be λ_x/λ_D times different. Hence, the time scale $(\lambda_t)_x$ that yields the displacement along x in the scale λ_x must be given by

$$(\lambda_t)_x = (\lambda_t)_s \frac{\lambda_x}{\lambda_D} \tag{6.81}$$

Substituting for $(\lambda_t)_s$ the values (6.64) we obtain

$$(\lambda_t)_x = \frac{\lambda_x}{\lambda_{v_*}} \tag{6.82}$$

and consequently

$$(\lambda_t)_x = \frac{\lambda_x^{\frac{3}{2}}}{\lambda_y} = \frac{\sqrt{(\lambda_y)}}{n^{\frac{3}{2}}} \tag{6.83}$$

Observe that for the conventional values of λ_y and n, the magnitudes of the time scales considered above are compared as follows:

$$(\lambda_t)_y < (\lambda_t)_x < \lambda_t < (\lambda_t)_s \tag{6.84}$$

The quickest is the formation of erosion and accretion, then the displacement of granular material from one location x to another, then the same for the fluid; the motion of individual grains being the slowest.

REFERENCES

1. A. Shields, *Anwendung der Ähnlichkeits mechanik und der Turbulenzforschung auf die Geschiebebewegung*, Mitteilungen der Preuss. Versuchs Anst. f. Wasserbau u. Schiffbau, Berlin, **26** (1936). See also translation by W. P. Ott and J. C. van Uchelen, U.S. Dept. Agriculture, Soil Conservation Service Coop Lab., Calif. Inst. Tech.
2. R. A. Bagnold, 'The Flow of Cohesionless Grains in Fluids', *Roy. Soc., London, Phil. Trans.*, Series A, No. 964, **249** (1956).
3. H. A. Einstein, 'Formulas for Transportation of Bed Load', *Transactions, A.S.C.E.*, **V.**, 107, pp. 561–573 (1942).
4. H. A. Einstein, *The Bed Load Function for Sediment Transportation in Open Channel Flows*, U S. Dept. of Agriculture, Soil Conservation Serv., Tech. Bull. 1026 (1950).
5. H. A. Einstein, 'River Sedimentation', Section 17-II in *Handbook of Applied Hydrology* (Ed. Ven Te Chow), McGraw-Hill (1964).
6. J. Rottner, 'A Formula for Bed Load Transportation', *La Houille Blanche*, **V,** 4, No. 3, pp. 301–307 (1959).
7. M. S. Yalin, 'An Expression for Bed-Load Transportation', *Proc. A.S.C.E.*, Jour. Hydr. Div., **89,** HY3 (1963).
8. M. S. Yalin, 'On the Average Velocity of Flow over a Mobile Bed', *La Houille Blanche*, **1,** Jan.–Feb. (1964).
9. M. S. Yalin, *Similarity in Sediment Transport by Currents*, Ministry of Technology, Hydr. Res. Paper No. **6,** H.M. Stationary Office (1965).
10. M. S. Yalin, 'Geometric Properties of Sand Waves', *Proc. A.S.C.E., Jour. Hydr. Div.*, **90,** HY5, Sept. (1964).
11. M. S. Yalin, 'Sand Waves'. Ministry of Technology, Hydr. Res., 1964 (Annual Rep.), H.M. Stationary Office, London, 1965.
12. A. J. Raudkivi, *Loose Boundary Hydraulics*, Pergamon Press (1967).

13. V. N. Goncharov, *Dynamics of Channel Flow*, Gidromet. Izdat., Leningrad (1962) (in Russian). English transl. by Israel Program of Scientific Translations (UNESCO) (1964).
14. M. A. Velikanov, *Dynamics of Alluvial Streams*, **II** (Sediment and Flow Bed) State Publishing House for Theor. and Techn. Literature, Moscow (1955) (in Russian).
15. H. A. Einstein, Ning Chien, 'Similarity of Distorted Models with Movable Beds', *Transactions A.S.C.E.* Paper 2805, **121** (1956).
16. M. S. Yalin, 'Über die dynamische Ähnlichkeit der Geschiebebewegungen', *Die Wasserwirtschaft*, **8/9,** August/September (1960).
17. F. Engelund, E. Hansen, *The Hydraulic Resistance of Alluvial Streams*, Techn. Univ. of Denmark, Costal Eng. Lab., Basic Research Progress Rep. **8,** 1965.
18. M. S. Yalin, *Mechanics of Sediment Transport*, Pergamon Press (In preparation).
19. E. W. Lane, E. W. Eden, 'Sand Waves in the Lower Mississippi River', *Journal of the Western Society of Engineers*, **45,** No. 6 (1940).

7 Waves

7.1 General

We now come to the study of similarity as applied to waves, certainly the most difficult subject encountered so far. When the application of the exact method of the theory of similarity becomes impossible, then, as has repeatedly been shown in the preceding chapters, a model is designed using mathematical expressions describing the relationships between the

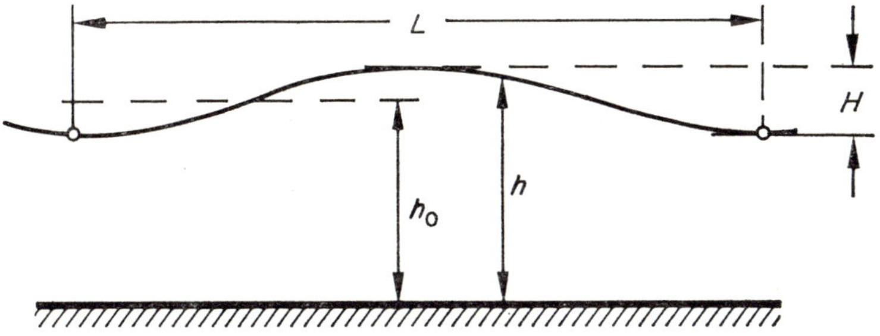

FIG 7.1

quantities involved. Unfortunately, an adequate mathematical theory exists only for wave motion of an ideal (inviscid) fluid. This theory, as is clear from its definition, cannot provide information on such quantities as shear stress, roughness, viscosity, boundary layer thickness and so on, which are of importance in the design of a model. Reliable information on the oscillatory boundary layer is available solely for viscous (laminar) fluid motion over a smooth plain bed. Considering that in almost all cases of civil engineering practice fluid motion is turbulent, while the bed is neither smooth nor plain, it will be clear how limited is the application of existing reliable methods when practical problems occur. It is small wonder then that, to date, no firm method has been established for the design of a small scale wave and/or tidal model where the influence of roughness, viscosity, shear stress, etc. is important (as in all models with mobile bed). Very often the design depends simply on experience and intuition rather than on any method at all.

Since the subject is of considerable importance and although the available tools have many limitations, I believe some attempt should be made in this book to discuss the similarity of model and prototype waves. Accordingly, in this chapter I have set out to develop such a discussion in the light of existing knowledge.

We begin with the classification of waves and the related terminology. Consider three geometric parameters of wave motion

wave length L
wave height H
still water depth h_o

Out of them we can form two dimensionless ratios

$$r_1 = \frac{h_o}{L}; \quad r_2 = \frac{H}{L} \tag{7.1}$$

Depending on the numerical orders of r_1 and r_2, waves can be classified as being of certain types. Geometrically, *deep water waves* should be distinguished from *shallow water waves* depending on whether r_1 is large or small. However, mechanically the wave motion is practically independent of the depth h_o already when r_1 is larger than $\approx 1/2$. In fact the amplitude of orbital motion decreases as a continuous function of the distance, measured downwards from still water level; no abrupt change takes place in the laws governing the properties of wave motion along the distance mentioned. Accordingly, the value $\approx 1/2$, for the exclusion of the influence of h_o, is a matter of convention only.[1] In keeping with that convention, for the following study of similarity (which is related to the mechanics of wave motion) the following definitions are adopted

$$\left.\begin{array}{ll} \text{deep water waves:} & \text{if } r_1 > \frac{1}{2} \\ \text{shallow water waves:} & \text{if } r_1 < \frac{1}{2} \end{array}\right\} \tag{7.2}$$

If the ratio r_1 is smaller than about 1/20, then it can be treated simply as $r_1 \to 0$. In this case the form of the expressions, i.e. the mechanical character of the wave motion, undergoes a substantial change, and therefore it would be reasonable to consider the waves corresponding to $r_1 < 1/20$ as a separate group. Accordingly, the waves can also be classified into *short* and *long* as follows:

$$\left.\begin{array}{ll} \text{short waves:} & \text{if } r_1 > 1/20 \\ \text{long waves:} & \text{if } 0 < r_1 < 1/20 \end{array}\right\} \tag{7.3}$$

The classification introduced above can be shown in the following unified manner[2,3,4].

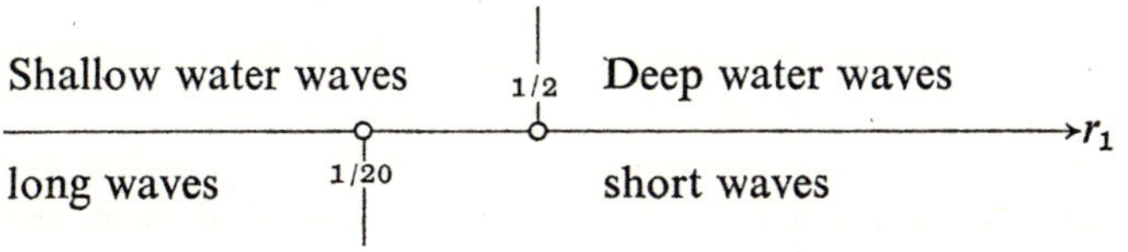

Clearly this scheme, which indicates how some rather vague adjectives should be associated with some values of r_1 (which are to a considerable extent arbitrary), cannot be unique, and the possibility of the introduction of other schemes cannot be overlooked. For example[5], the following scheme is adopted

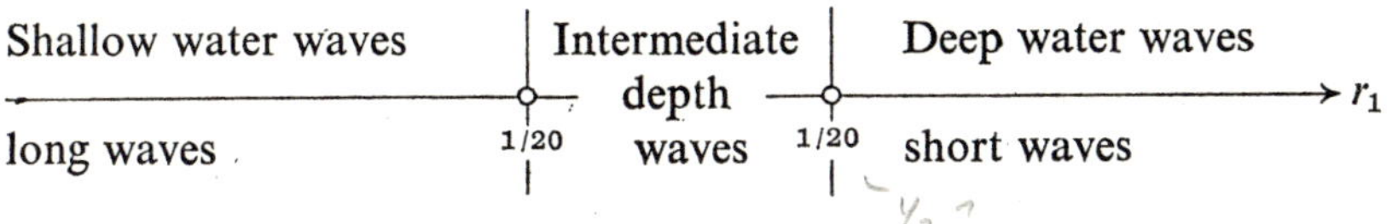

However, to maintain consistency in terminology, in the present chapter only the former scheme will be used.

Let us now consider the ratio r_2. The mathematical expressions are greatly simplified if r_2 can be treated as small (i.e. if any $r_2{}^m$, where m is an integer ($m > 1$), can be neglected in comparison to r_2). We therefore separate the waves corresponding to small r_2 (and thus to the simplest version of mathematical expressions) and introduce the following terminology:

$$\left.\begin{array}{l}\text{waves with small amplitude (wave height):}\\ \qquad\qquad\text{if } r_2 \text{ is treated as infinitesimal}\\ \text{waves with finite amplitude (wave height):}\\ \qquad\qquad\text{if } r_2 \text{ is treated as finite}\end{array}\right\} \tag{7.4}$$

In the following study, when dealing with short waves we will assume automatically that we are dealing with progressive gravity waves which are generated by wind (*wind waves*), ship motion, etc. and which have the period T of the order of several seconds (say 5 to 15). Furthermore, for the conditions usually met with in practice (water depth, viscosity, roughness, etc.), we will assume that the thickness δ of the oscillatory boundary layer is much less than the water depth h_o (that $\delta \ll h_o$). When dealing with long waves, we will be dealing with *tidal waves* having a period $T \approx 12$ hours. In this case it will invariably be assumed that δ, which increases (directly or indirectly) with T, has reached the ceiling h_o (i.e. that $\delta = h_o$ is valid).

7.2 Deep Water Waves $\left(\frac{h_o}{L} = r_1 > \frac{1}{2} \right)$

From the definitions above, it is clear that in the case of deep water waves the water depth h_o cannot be a characteristic parameter (Fig. 7.2). Furthermore, if consideration is restricted to the usual gravity waves mentioned, then water can be treated as an ideal fluid ($\mu = 0$). Hence, deep water waves can be completely defined by the fluid density ρ, the acceleration due to gravity g, the wave height H and the wave length L. Accordingly, for a specific shape of waves (sinusoidal waves, trochoidal waves, etc.)

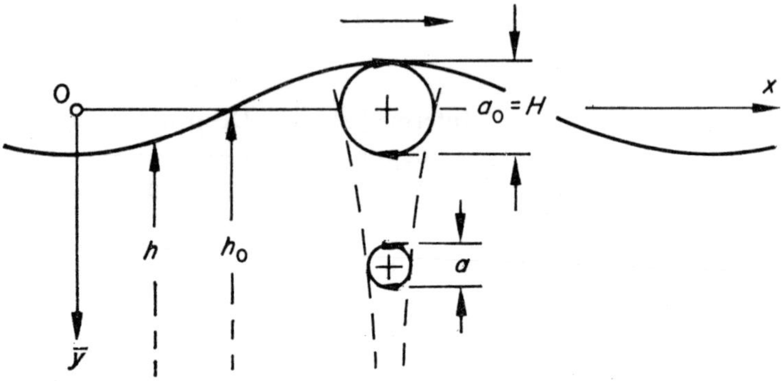

FIG. 7.2

the phenomenon can be described by the following four characteristic parameters

$$\rho, g, H, L \tag{7.5}$$

Selecting ρ, g and L as basic quantities we arrive at the single ($4 - 3 = 1$) dimensionless variable

$$X_1 = \rho^0 g^0 L^{-1} H = \frac{H}{L} = r_2 \tag{7.6}$$

and at the following dimensionless version of a property A

$$\Pi_A = \rho^{x_A} g^{y_A} L^{z_A} A \tag{7.7}$$

which are related to each other by

$$\Pi_A = \varphi_A(r_2) \tag{7.8}$$

For example, if A is *the wave period* T, then

$$\Pi_T = \rho^0 g^{\frac{1}{2}} L^{-\frac{1}{2}} T = T \sqrt{\frac{g}{L}} \tag{7.9}$$

$r_1 = \frac{h_0}{L}$ $r_2 = \frac{H}{L}$

and thus

$$T\sqrt{\frac{g}{L}} = \varphi_T(r_2) \tag{7.10}$$

or

$$T = \varphi_T(r_2)\sqrt{\frac{L}{g}} \tag{7.11}$$

which is indeed so, as is known from the wave theory. In particular if the wave is of small amplitude, i.e. if $r_2 \to 0$, then

$$\varphi_T(0) = \text{const} \tag{7.12}$$

and thus

$$T = \text{const}\sqrt{\frac{L}{g}} \tag{7.13}$$

Comparing (7.13) with the well-known expression from Airy's theory of waves

$$T = \sqrt{2\pi}\sqrt{\frac{L}{g}} \tag{7.14}$$

it will be seen that the expression supplied by the theory of dimensions is complete up to the numerical value of the constant (const $= 2\pi$).

If a property A varies as a function of position x, $\bar{y}$ and the time t (Fig. 7.2), then the dimensionless version of such a property A can be expressed as follows

$$\Pi_A = \varphi_A(r_2, \xi, \bar{\eta}, \theta) \tag{7.15}$$

where ξ, $\bar{\eta}$, are the dimensionless coordinates of the position while θ is the dimensionless time. Bearing in mind that the basic quantities are ρ, g and L, we have for ξ, $\bar{\eta}$ and θ

$$\xi = \frac{x}{L}, \quad \bar{\eta} = \frac{\bar{y}}{L}, \quad \theta = t\sqrt{\frac{g}{L}} \tag{7.16}$$

Consider, for example, as A, the orbit length a which varies with $\bar{y}$. We have

$$\Pi_a = \rho^0, g^0 L^{-1} a = \frac{a}{L} \tag{7.17}$$

and thus

$$\frac{a}{L} = \varphi_a(r_2, \bar{\eta}) \tag{7.18}$$

$r_1 = \frac{h_o}{L}$ $r_2 = \frac{H}{L}$

From the theory of waves, it is known that

$$\frac{a}{H} = \mathrm{e}^{2\pi \frac{y}{L}} \tag{7.19}$$

Thus, the value of a is given by a function of two variables φ_a, which has the form

$$\varphi_a(r_2, \bar{\eta}) = r_2 \mathrm{e}^{2\pi\bar{\eta}} \tag{7.20}$$

Let A be the orbital velocity U. We have

$$\Pi_U = \rho^0 g^{-\frac{1}{2}} L^{-\frac{1}{2}} U = \frac{U}{\sqrt{gl}} \tag{7.21}$$

which is a Froude number. Clearly U must also vary with r_2 and $\bar{y}$ and thus

$$\frac{U}{\sqrt{gL}} = \varphi_U(r_2, \bar{\eta}) \tag{7.22}$$

must be valid. On the other hand, the orbital velocity, according to definition, must be given by

$$U = \frac{2a}{T} \tag{7.23}$$

Substituting here, instead of T and a, their values (7.11) and (7.18), we arrive at

$$U = \frac{2\varphi_a(r_2, \bar{\eta})L}{\sqrt{\left(\frac{L}{g}\right)}\varphi_T(r_2)} = 2\sqrt{(gL)}\frac{\varphi_a(r_2, \bar{\eta})}{\varphi_T(r_2)} \tag{7.24}$$

i.e.

$$\frac{U}{\sqrt{gL}} = 2\frac{\varphi_a(r_2, \bar{\eta})}{\varphi_T(r_2)} \tag{7.25}$$

The eqn. (7.25) shows clearly that the function φ_U is but a symbol which stands for the double of the ratio of the functions φ_a and φ_T. In the present chapter, it is relevant to remember that

$$a, T \text{ and } U$$

are *interdependent properties* and thus only two of them can be used (as parameters) for the description of a quantity.

Let us now consider the determination of scales. In realistic models the scales of two, out of four parameters (7.5), are equal to unity

$$\lambda_\rho = 1, \lambda_g = 1 \tag{7.26}$$

$r_1 = \frac{h_o}{L}$ $r_2 = \frac{H}{L}$

Thus, the realization of a dynamically similar small scale model is possible, for the third scale λ_L can be selected less than unity. We have

$$\lambda_{X_1} = \lambda_{r_2} = 1 \tag{7.27}$$

i.e.

$$\lambda_L = \lambda_H \tag{7.28}$$

which confirms that the dynamically similar model must be undistorted. Denoting the length scale simply by λ and considering (7.26) we arrive at the following expression of the scale λ_A of a property A of the wave motion under consideration

$$\lambda_{\Pi_A} = \lambda^{z_A} \lambda_A = 1 \tag{7.29}$$

i.e.

$$\lambda_A = \lambda^{-Z_A} \tag{7.30}$$

For example:

$$\left.\begin{array}{llll} \text{when } A = T & \text{then } Z_A = -\frac{1}{2} & \text{and thus} & \lambda_T = \lambda^{\frac{1}{2}} \\ \text{when } A = a & \text{then } Z_A = -1 & \text{and thus} & \lambda_a = \lambda \\ \text{when } A = U & \text{then } Z_A = -\frac{1}{2} & \text{and thus} & \lambda_U = \lambda^{\frac{1}{2}} \\ & & & \ldots \text{etc.} \end{array}\right\} \tag{7.31}$$

In the derivations above, no condition was imposed on the magnitude of the ratio r_2. Thus the scale relations determined above are equally valid for waves of small and finite amplitude.

Shallow Waves.

7.3 Short (Wind) Waves Over a Rigid Bed $\left(r_1 = \frac{h_o}{L} > \frac{1}{20}\right)$

7.3.1 General Usually it is assumed that if the additional parameter h_o (the water depth) is introduced, i.e. if the set (7.5) is generalized into

$$\rho, g, H, L, h_o \tag{7.32}$$

and (7.8) into

$$\Pi_A = \varphi_A \left(\frac{h_o}{L}, \frac{H}{L}\right) = \varphi_A(r_1, r_2) \tag{7.33}$$

then the transition from deep water waves to shallow water waves is achieved. Indeed, in the conventional treatment the expressions for shallow and deep water differ only because of the ratio h_o/L. Furthermore, the shallow water wave relations such as (7.33) or

$$\Pi_A = \varphi_A(r_1, r_2, \xi, \eta, \theta) \tag{7.34}$$

are treated as if they are valid throughout the region

$$0 < y < h_o \tag{7.35}$$

Clearly, such a treatment is misleading. It should be clearly pointed out that the relationships (7.33) and (7.34) are valid only in the upper part (*outer region*)

$$\delta < y < h_o \tag{7.36}$$

of the region (7.35) (see Fig. 7.3): the fluid motion in the lower part

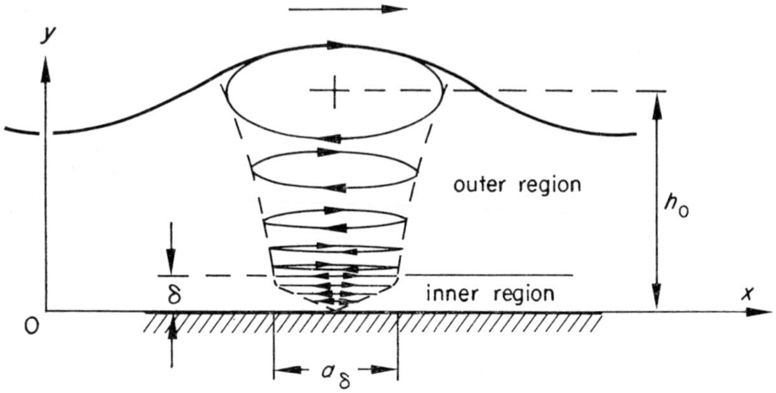

FIG. 7.3

(*inner region*) $$0 \text{ (or } k_s) < y < \delta \tag{7.37}$$

being governed by entirely different laws (where δ is the thickness of the boundary layer). If the fluid motion due to waves is to be considered for the whole region (7.35), then the introduction of h_o alone is not sufficient, for the fluid motion in the vicinity of a flow boundary cannot be represented by that of an (inviscid) ideal fluid. It follows that the description of the phenomenon can be complete only if the viscosity μ and the roughness k_s are also introduced. In other words, the transition from deep water waves to shallow water waves requires the introduction of three additional characteristic parameters h_o, μ and k_s. Accordingly, the set of characteristic parameters of the wave motion in shallow waters, in general, consists of seven independent quantities

$$\rho, g, H, L, h_o, \mu, k_s \tag{7.38}$$

Using, for instance, ρ, g and L again as basic quantities, we arrive at the following $7 - 3 = 4$ dimensionless variables

$$\left.\begin{aligned} X_1 &= \frac{H}{L} = r_2 \\ X_2 &= \frac{h_o}{L} = r_1 \\ X_3 &= \frac{k_s}{L} \\ X_4 &= \frac{L\sqrt{Lg}}{\nu} \end{aligned}\right\} \tag{7.39}$$

In a realistic model

$$\lambda_\rho = 1,\ \lambda_g = 1,\ \lambda_\mu = 1 \tag{7.40}$$

i.e. the scales of three dimensionally independent parameters are equal to unity. Thus the realization of a small scale dynamically similar model is, in general, impossible. It is possible only in certain special cases. However, before considering such cases, it is advisable first to pause and consider some aspects of the boundary layers of the oscillatory fluid motions generated by short (wind) waves.

7.3.2 Oscillatory Boundary Layer due to Short Waves The boundary layer thickness δ is defined as the value of y which forms the borderline between the outer region (where the shear stress τ can be assumed to be zero) and the inner region (where τ is different from zero). The orbit length a_δ and the orbital velocity U_δ corresponding to $y = \delta$, belong to both outer and inner flows, i.e. they can be regarded as properties of the outer region, and thus they can be given by parameters describing the wave motion. On the other hand the 'border quantities' a_δ and U_δ can be used as characteristic parameters for the definition of fluid motion within the boundary layer. In fact, the kinematic state of the boundary layer flow can be defined by any two out of three quantities a_δ, U_δ and T (see the text between the eqns. (7.25) and (7.26)). Choosing a_δ and T to describe the kinematics we can define the boundary layer flow by the following five dimensional parameters

$$a_\delta,\ T,\ \rho,\ \mu,\ k_s \tag{7.41}$$

The parameters a_δ, T and ρ have independent dimensions and (unlike μ and k_s) their presence in the set above does not depend on the regime of the boundary layer flow. Selecting them as basic quantities we arrive at the following dimensionless variables

$$\frac{\nu T}{a_\delta^2}, \quad \frac{k_s}{a_\delta} \tag{7.42}$$

The dimensionless version of any property of the boundary layer flow, and thus the boundary layer thickness δ itself, must be a certain function of the two dimensionless variables above. Hence

$$\frac{\delta}{a_\delta} = \varphi_\delta \left(\frac{\nu T}{a_\delta^2}, \frac{k_s}{a_\delta} \right) \tag{7.43}$$

Unfortunately, no experimental family of the curves has been produced so far, that would extend our knowledge on the form of the function of these two variables. The experimental determination of the form φ_δ is relatively simple. For certain $(k_s)_i$, $(h_o)_i$ and H (and thus $(k_s/a_\delta)_i$) vary T (and thus $\nu T/a_\delta^2$) and measure δ (and thus δ/a_δ). Alter i and repeat the procedure. The experimental curve corresponding to each $(k_s/a_\delta)_i$ is a member of the family implied by (7.43). Apparently progress on this score has been impeded because δ was regarded as a function of five (dimensional) parameters (7.41) rather than of only 2 (as implied by (7.43)). All that is known for certain about the function φ_δ is its limiting form corresponding to viscous fluid motion.

In the vicinity of the bed, the influence of the pressure gradient and of convective acceleration, on the mechanical character of the fluid motion, is negligible[6,7,8]. In other words, one can assume that the friction force (acting per unit mass of the fluid) is brought into equilibrium by the acceleration (parallel to the stream lines) alone, and thus that the fluid motion is governed by the following differential equation

$$\frac{\partial u}{\partial t} = \frac{1}{\rho} \frac{\partial \tau}{\partial y} \tag{7.44}$$

Clearly, the absence of convective terms implies automatically that the flow within the boundary layer is parallel to the bed and consequently that δ is small in comparison to h_o (as is usually the case with short (wind) waves in shallow waters.

The differential equation (7.44) implies that any force acting on the fluid is negligible in comparison to $\partial\tau/\partial y$. Clearly, this can only be the case in the region $0 < y < \delta$, i.e. in the region of existence of the shear stress τ. Accordingly, the scale of τ can be associated with the length δ; while the dimensionless ordinate must be y/δ. Yet, very often one introduces y/h as the dimensionless ordinate, or (which is the same) one integrates the differential equation along [0; h] (instead of [0, δ]). Consequently, one determines the scale of τ in terms of the scale of the water depth h_0, although h_0 has no direct bearing on the motion within the boundary layer[9].* Such an approach is permissible only if $\lambda_\delta = \lambda_h$ or $\delta = h_0$ (i.e. long waves). Clearly, with a rough bed the lower limit is k_s instead of 0 in all the intervals mentioned above. Introducing

$$\eta = \frac{y}{\delta} \quad \text{and} \quad \theta = \frac{t}{T} \tag{7.45}$$

as the dimensionless variables and

$$\varphi_u = \frac{u}{u_\delta} \quad \text{and} \quad \varphi_\tau = \frac{\tau}{\tau_0} \tag{7.46}$$

as the dimensionless functions, where u_δ and τ_0 correspond to the *same* typical fraction θ of the period, we can express the eqn. (7.44) in the dimensionless form as follows

$$\frac{\partial \varphi_u}{\partial \theta} = K \frac{\partial \varphi_\tau}{\partial \eta} \qquad \left(\text{where } K = \left[\frac{u_\delta T}{\delta}\right]\left[\frac{\tau_0}{\rho u_\delta^2}\right]\right) \tag{7.47}$$

The solution of the equation above depends on the numerical value of the dimensionless combination K. It seems that *if* δ is much smaller than h_0 ($\delta \ll h_0$), then, for the given ρ, T, u_δ and τ_0 the thickness δ of the oscillatory boundary layer establishes itself so as to yield always the same constant value for the combination K. Indeed, as will be seen later, all the known expressions of δ (corresponding to both viscous and rough turbulent flows) can be obtained only if it is assumed that K is a constant

$$K = \left[\frac{u_\delta T}{\delta}\right]\left[\frac{\tau_0}{\rho u_\delta^2}\right] = \text{const} \tag{7.48}$$

The value of the shear stress τ_0 produced by the oscillatory fluid motion generated by progressive waves can be expressed as follows[8]

$$\tau_0 = \alpha_\tau \rho u_\delta^2 + \beta_\tau \gamma \delta \bar{S} \tag{7.49}$$

Here the quantities τ_0, u_δ and $\bar{S} = -\partial h/\partial x$ (the slope of the free surface) correspond to the same section and instant.

(i) If the oscillatory motion is viscous, then, by a rigorous analysis[8]

$$\alpha_\tau = \frac{(\text{const})_1}{\left(\dfrac{U_\delta \delta}{\nu}\right)}, \quad \beta_\tau = \text{const}_\beta \tag{7.50}$$

where U_δ is the mean orbital velocity at the level $y = \delta$.

* The depth h_0 contributes, in conjunction with H, to the formation of a_δ, and so only bears indirectly (by means of a_δ) on the motion within the boundary layer.

(ii) If the oscillatory motion is rough turbulent, then, as has been found from the evaluation of experimental data (in Ref. 8) a very satisfactory agreement is achieved if α_τ is expressed according to the logarithmic velocity distribution, while β_τ is treated as a function of relative roughness alone.

$$\alpha_\tau = \frac{1}{\left(2{\cdot}5 \ln 30{\cdot}1 \dfrac{\delta}{k_s}\right)^2}, \quad \beta_\tau = \varphi_B \left(\frac{k_s}{\delta}\right) \tag{7.51}$$

The experiments were performed using long waves where δ was equal to the flow depth h_o. Hence, the coefficient α_τ was in fact

$$\alpha_\tau = \frac{1}{\left(2{\cdot}5 \ln 11 \dfrac{h_o}{k_s}\right)^2}$$

which implies $\alpha_\tau = 1/c^2$. Since, in the case of the short waves under consideration, δ is much smaller than h_o, the expression of α_τ has been modified accordingly. In (7.51) the quantity α_τ is expressed so as to imply that the value of τ_o corresponding to the instant when

$$u_\delta = (u_\delta)_{max} \quad \text{while} \quad \bar{S} = 0$$

is the same as that of a unidirectional flow having a logarithmic velocity distribution (and which possesses at the level $y = \delta$ a velocity u that is identical to $u_\delta = (u_\delta)_{max}$). The consideration of logarithmic velocity distribution for the oscillatory boundary layer is in accordance with the contemporary convention[10,11]. However[8], strictly speaking, this can be correct only in the lower parts of the boundary layer for, when approaching $y = \delta$, the rate of change du/dy must approach zero; yet the derivative of a logarithmic function can never be zero for a finite value of its variable.

The relation (7.49) has been subjected to the following criticism: 'it is unlikely that the slope $\bar{S}$ of the free surface can *generate* the shear stress acting on the bed'. The invalidity of such a criticism lies in regarding 'generated by . . .' and 'expressible by . . .' as the same thing. Indeed, let us assume that A generates *only* B ($B = f(A)$), while C is caused *only* by B ($C = \varphi(B)$). Clearly, in this case A is not a direct cause of C, yet one can always avoid B completely and express C as a function of A ($C = \varphi(f(A)) = \psi(A)$). A particular combination of $\bar{S}$ and u_δ causes *only* a particular field of the velocities u (in the vicinity of the bed) the value of τ_o being the consequence of that field. The possibility for the form

(7.49) to be valid also for the (disputable) rough turbulent flow can be explained as follows: Excluding μ from the set (7.38) we arrive at

$$\rho,\ g,\ H,\ L,\ h_o,\ k_s \tag{7.52}$$

Any time-dependent property of the rough turbulent wave motion must be a certain function of the above parameters and of the dimensionless instant θ. Thus

$$\begin{aligned} \tau_o &= f_1(\rho, g, H, L, h_o, k_s, \theta) \\ u_\delta &= f_2(\rho, g, H, L, h_o; k_s, \theta) \\ \bar{S} &= f_3(\rho, g, H, L, h_o, k_s, \theta) \\ \delta &= f_4(\rho, g, H, L, h_o, k_s, \theta) \end{aligned}$$

eliminating three quantities H, L, h_o from four equations above, we arrive at

$$\tau_o = f_5(\rho, g, u_\delta, \bar{S}, \delta, k_s, \theta) \tag{7.53}$$

i.e. at the possibility of expressing τ_o by u_δ and $\bar{S}$. In fact, with the exception of θ, the variables forming (7.49) are precisely those of the expression above (take into account (7.51)). The presence of θ in the expression of τ_o above implies that the consideration of the instantaneous values of u_δ and $\bar{S}$ might not be sufficient, or that the form of the functional relationship (such as (7.49)) where a separate θ is not present, might vary from one instant to another. However, such a possibility certainly does not apply to the equation under consideration, for, as can be seen from the series of diagrams published in Ref. [8] (plate No. 2 to 24) all the τ_o-curves computed by using the measured values of u_δ and $\bar{S}$, and by utilizing always the same form of (7.49), agree very well with the recorded oscillogram of τ_o throughout the periodic cycle ($0 < \theta < 1$). It has to be pointed out, however, that the explanations above justify only the adequacy of the parameters used, and consequently the validity of the form (7.49). It is also evident that in the case of a rough turbulent motion, the dimensionless coefficients α_τ and β_τ must be functions of the dimensionless roughness δ/k_s. But whether in the case of short waves the form of the function α_τ is indeed as implied by (7.51) remains a question to be answered in the future.

For the sake of simplicity, the thickness of the oscillatory boundary layer can be defined as that value of δ which corresponds to the instant when

$$u_\delta = (u_\delta)_{max} \quad \text{and} \quad \bar{S} = 0 \tag{7.54}$$

That is, when the shear stress acting on the bed is given, according to (7.49), by the following expression

$$\tau_o = \bar{\tau}_o = \alpha_\tau \rho (u_\delta)_{max}{}^2 \tag{7.55}$$

(i) If the oscillatory motion of the fluid in the vicinity of the bed is viscous, then α_τ is given by (7.50). Thus

$$\frac{\bar{\tau}_0}{\rho(u_\delta)_{max}^2} = \frac{(\text{const})_1}{\left(\dfrac{U_\delta \delta}{\nu}\right)} \tag{7.56}$$

Substituting this value in (7.48) we arrive at

$$\frac{\dfrac{(u_\delta)_{max} T}{\delta}}{\dfrac{U_\delta \delta}{\nu}} = \frac{\text{const}}{(\text{const})_1} \tag{7.57}$$

and consequently at the well-known formula

$$\delta = \text{const}_1 \sqrt{T\nu} \tag{7.58}$$

$$\left(\text{where: const}_1 = \left[\frac{\text{const}}{(\text{const})_1} \frac{U_\delta}{(u_\delta)_{max}}\right]^{-\frac{1}{2}}\right)^*$$

The relation (7.58) can, of course, be derived in a considerably shorter way. But the intention here is to introduce a unified method that is applicable to both viscous and turbulent boundary layers.

(ii) If the oscillatory motion is rough turbulent, then α_τ is assumed to be given by (7.51). Thus

$$\frac{\bar{\tau}_0}{\rho(u_\delta)_{max}^2} = \frac{1}{\left(2{\cdot}5 \ln\left(30{\cdot}1 \dfrac{\delta}{k_s}\right)\right)^2} \tag{7.59}$$

Substituting this value in (7.48) we obtain

$$\frac{(u_\delta)_{max} T}{\delta} = \text{const}\left(2{\cdot}5 \ln 30{\cdot}1 \frac{\delta}{k_s}\right)^2 \tag{7.60}$$

i.e.

$$\frac{a_\delta}{\delta} = \text{const}_2 \left(2{\cdot}5 \ln 30{\cdot}1 \frac{\delta}{k_s}\right)^2 \tag{7.61}$$

$$\left(\text{where: const}_2 = \frac{1}{2} \text{const} \frac{U_\delta}{(u_\delta)_{max}}\right)^{**}$$

* For the specified shape of the curve representing the variation of $u\delta$ with t the average (orbital) velocity U_δ is related to the maximum velocity $(u_\delta)_{max}$ by a constant proportion.

** ibid.

which has exactly the same form as the expression derived in Ref. 10, using a different mathematical approach.

Clearly, the derived forms (7.58) and (7.61) are two special cases of the general form (7.43) determined from the dimensional considerations. Indeed, for case (i) the relationship (7.43) reduces

$$\frac{\delta}{a_\delta} = \bar{\varphi}_\delta \left(\frac{\nu T}{{a_\delta}^2}\right) \tag{7.62}$$

while for case (ii) it becomes

$$\frac{\delta}{a_\delta} = \bar{\bar{\varphi}}_\delta \left(\frac{k_s}{a_\delta}\right) = \bar{\bar{\psi}}_\delta \left(\frac{k_s}{\delta}\frac{\delta}{a_\delta}\right) \tag{7.63}$$

or

$$\frac{a_\delta}{\delta} = \psi_\delta \left(\frac{\delta}{k_s}\right) \tag{7.64}$$

Observe that the derived expressions (7.58) and (7.61) are in agreement with the functions (7.62) and (7.64) determined from dimensional considerations. Indeed, they indicate that the functions $\bar{\varphi}_\delta$ and $\bar{\bar{\psi}}_\delta$ are of the following form

$$\bar{\varphi}_\delta \left(\frac{\nu T}{{a_\delta}^2}\right) = \text{const}_1 \sqrt{\left(\frac{\nu T}{{a_\delta}^2}\right)} \tag{7.65}$$

and

$$\bar{\bar{\psi}}_\delta \left(\frac{\delta}{k_s}\right) = \text{const}_2 \left(2{\cdot}5 \ln 30{\cdot}1 \frac{\delta}{k_s}\right)^2 \tag{7.66}$$

The logarithmic function is rather cumbersome to use for determining scales. From Fig. 7.4 it is clear that this function can be approximated by the far more convenient power law as

$$2{\cdot}5 \ln 30{\cdot}1 \frac{\delta}{k_s} \approx \overline{\text{const}} \left(\frac{\delta}{k_s}\right)^\alpha \tag{7.67}$$

throughout the practical range (see Fig. 7.4)

$$\approx 2k_s < \delta < \approx 100k_s \tag{7.68}$$

if the value of α is selected as

$$\alpha = \frac{1}{6} \tag{7.69}$$

Accordingly, the relationship (7.61) can be replaced by

$$\frac{a_\delta}{\delta} = \text{const} \left(\frac{\delta}{k_s}\right)^{\frac{1}{3}} \tag{7.70}$$

which gives at once

$$\delta = \text{const}\, a_\delta^{\frac{3}{4}} k_s^{\frac{1}{4}} \tag{7.71}$$

Using (7.58) and (7.70) in the equation above, we arrive at the following expressions of the boundary layer thickness scale

$$\lambda_\delta = \lambda_T^{\frac{1}{2}} \lambda_\nu^{\frac{1}{2}} \quad \text{(viscous fluid)} \tag{7.72}$$

$$\lambda_\delta = \lambda_a^{\frac{3}{4}} \lambda_{k_s}^{\frac{1}{4}} \quad \text{(rough turbulent)} \tag{7.73}$$

The exponents in the scale relationship (7.73) approximate to the truth only in the range of δ values given by (7.68). Otherwise the value of the

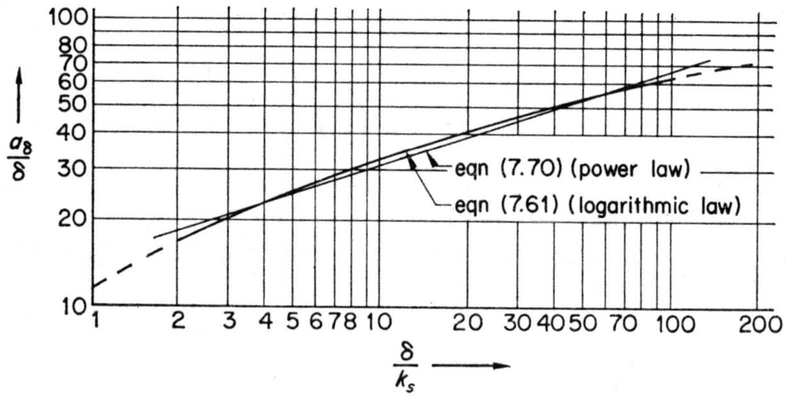

FIG. 7.4

exponent α, in (7.67), and thus the values of the exponents in (7.73), should be modified accordingly. This can be done without knowing the value of δ as follows. Write (7.61) as

$$\frac{a_\delta}{k_s} = \frac{\delta}{k_s}\,\text{const}_2 \left(2{\cdot}5 \ln 30{\cdot}1 \frac{\delta}{k_s}\right)^2 \tag{7.61$'$}$$

and plot the curve representing this relationship between a_δ/k_s and δ/k_s in a log log system of coordinates. Knowing a_δ/k_s one can determine from this curve δ/k_s and a straight line may be drawn representing the curve in the region of the values in question. Using this straight line implying

$$\frac{a_\delta}{k_s} = \text{const} \left(\frac{\delta}{k_s}\right)^\varepsilon$$

one arrives at the scale relationship

$$\lambda_\delta = \lambda_a^{1/\varepsilon} \lambda_{k_s}^{1-1/\varepsilon}$$

where ε is the slope of the straight line.

7.3.3 Model Scales Since the fluid motion in the outer region

$$\delta \leqslant y < h_o$$

can be represented by that of an ideal (frictionless) fluid, it cannot be dependent on the parameters μ and k_s, and thus on the dimensionless variables X_3 and X_4, which are dependent on μ and k_s (see eqn. (7.39)). Thus the dynamic similarity of the fluid motion in the outer region is given by

$$\lambda_{X_1} = 1 \quad \text{and} \quad \lambda_{X_2} = 1 \tag{7.74}$$

that is

$$\lambda_H = \lambda_h = \lambda_L \tag{7.75}$$

that is by the geometric similarity alone. Let λ be the geometric model scale. Since the basic quantities are ρ, g and L the scale of any property A of the motion in the outer region is given by

$$\lambda_{\Pi_A} = \lambda_\rho^{x_A} \lambda_g^{y_A} \lambda_L^{z_A} \lambda_A = 1 \tag{7.76}$$

i.e.

$$\lambda_A = \lambda^{-z_A} \tag{7.77}$$

since $\lambda_\rho = \lambda_g = 1$ while $\lambda_L = \lambda$. Observe, that (7.77) is the same as (7.31), which means that the expressions (7.31) of deep water waves are also valid for the outer region of short waves (in shallow waters).

The quantities a_δ and u_δ correspond to the borderline between the outer and inner regions. Thus they can be given by the scale relationships of the outer region and consequently we can write

$$\lambda_{a_\delta} = \lambda \quad \text{and} \quad \lambda_{u_\delta} = \sqrt{\lambda} \tag{7.78}$$

Let us now substitute these values in the relationship

$$\frac{\lambda_{a_\delta}}{\lambda_\delta} \frac{\lambda_\tau}{\lambda_{u_\delta}{}^2} = 1 \tag{7.79}$$

which follows at once from the expression (7.48). We obtain

$$\lambda_\tau = \lambda_\delta \tag{7.80}$$

Thus in the case of oscillatory motion, where $\delta \ll h_o$, the scale of the shear stress coincides with the scale of the boundary layer thickness. Substituting (7.73) and (7.72) in (7.80), and taking into account that $\lambda_\nu = 1$ while $\lambda_{a_s} = \lambda$, we arrive at the following values of the shear stress scale, which will frequently be used later on

$$\lambda_\tau = \lambda_T{}^{\frac{1}{2}} \quad \text{(viscous)} \tag{7.81}$$

$$\lambda_\tau = \lambda^{\frac{3}{4}} \lambda_{k_s}{}^{\frac{1}{4}} \quad \text{(rough turbulent)} \tag{7.82}$$

Even though the motion in the outer region is dynamically similar, the motion, as a whole, cannot be such for we are dealing with a small scale model ($\lambda < 1$) and yet the scales of three dimensionally independent parameters ρ, g and μ are equal to unity (as implied by (7.40)). The question arises 'what is then wrong with the small scale model under consideration?' Let us attempt to answer this question.

Consider first the viscous fluid motion. Since $\lambda_\nu = 1$, the relation (7.72) gives

$$\lambda_\delta = \lambda_T^{\frac{1}{2}} \tag{7.83}$$

On the other hand, the scale of the period for the undistorted model under consideration is given by

$$\lambda_T = \lambda^{\frac{1}{2}} \tag{7.84}$$

Indeed, the period T is the property of both inner and outer regions, and thus its scale can obviously be given by (7.31). Equating (7.83) and (7.84) we obtain

$$\lambda_\delta = \lambda^{\frac{1}{4}} \tag{7.85}$$

Thus the 'wrongness' is due to the fact that the scale of the 'length' δ is not equal to the geometric scale λ as it should be in a dynamically similar model. Since $\lambda < 1$, it is clear that

$$\lambda^{\frac{1}{4}} > \lambda \tag{7.86}$$

and thus, that the model boundary layer is (relatively) thicker than the prototype boundary layer. Clearly, a similar situation will also be present for turbulent flows so long as they are dependent on viscosity in the vicinity of the bed (hydraulically smooth and transitional turbulent flows). If, however, the turbulent flow is fully developed (rough turbulent), then μ is no longer a characteristic parameter, even as far as the vicinity of the bed is concerned. In this case, we have the scales of only two parameters (ρ and g) that are equal to unity, and thus dynamic similarity for the motion as a whole must be possible. Indeed, excluding X_4 (reflecting the influence of μ) and equating the scales of the remaining variables X_1, X_2 and X_3 to unity, we arrive at

$$\lambda_\rho = \lambda_{h_0} = \lambda_{k_s} = \lambda_L = \ldots = \lambda \tag{7.87}$$

which implies that the dynamic similarity, of the motion as a whole, is given by the geometric similarity alone (including the similarity of roughness). For example, in this case there is no longer any discrepancy between λ_δ and λ. Indeed, the relation (7.73) gives

$$\lambda_\delta = \lambda^{\frac{3}{4}} \lambda_{k_s}^{\frac{1}{4}} \tag{7.88}$$

and since

$$\lambda_{k_s} = \lambda$$

we have

$$\lambda_\delta = \lambda \tag{7.89}$$

From (7.88) it is clear that if λ_{k_s} is *not* equal to λ, then the model boundary layer will be (relatively) thinner or thicker than the prototype boundary layer, depending on whether λ_{k_s} is smaller or larger than λ. However, the error due to the inequality $\lambda_{k_s} \neq \lambda$ will be damped considerably by the relatively small exponent 1/4 (e.g. if $\lambda_{k_s} = 2\lambda$, then λ_δ is only 1.19 times larger than λ).

The flow in the boundary layer, in general, has two different scales

vertical scale λ_δ
horizontal scale λ_{a_s}

that is, the flow within the boundary layer is *distorted.* Hence, for a property of the boundary layer motion in the vertical direction we have

$$\Pi_A = \rho^{x_A} T^{y_A} \delta^{z_A} A$$

and thus

$$\lambda_A = \lambda_\rho^{-x_A} \lambda_T^{-y_A} \lambda_\delta^{-z_A}$$

i.e.

$$\lambda_A = \lambda^{-\frac{y_A}{2}} \lambda_\delta^{-z_A} \tag{7.90}$$

For the property in the horizontal direction we have

$$\Pi_A = \rho^{x_A} T^{y_A} a_\delta^{z_A} A$$

and consequently

$$\lambda_A = \lambda_\delta^{-x_A} \lambda_T^{-y_A} \lambda_{a_\delta}^{-z_A}$$

i.e.

$$\lambda_A = \lambda^{-\left(z_A + \frac{y_A}{2}\right)} \tag{7.91}$$

Clearly, in the expression (7.90) one must substitute that value of λ_δ which corresponds to the regime under consideration.

7.3.4 Remarks

(i) No restriction was imposed on the numerical value of the ratio r_2 during the derivations in this section. Thus the relations derived are equally valid for wind waves of both small and finite amplitude.

(ii) Let us for brevity denote the combinations $\nu T/a_\delta^2$ and k_s/a_δ (see (7.42) by X_ν and X_k respectively. The possible regimes of the motion in oscillatory boundary layer, and the corresponding characteristic combinations, are

1	X_ν		viscous (laminar) regime	
2	X_ν		hydraulically smooth	turbulent regime
3	X_ν	X_k	transitional	turbulent regime
4		X_k	rough (fully developed)	turbulent regime

The transition 1 → 2 is given by a certain (critical) value of X_ν*; each of the transitions 2 → 3 and 3 → 4 being determined by a certain (critical) relation between X_ν and X_k. Clearly, if turbulent motion is not fully developed, then the value of δ/k_s can no longer be given (by (7.61)′) as a function of X_k alone, while the value of δ, and consequently of λ_δ, can no longer be determined by relations that are independent of $\mu \sim X_\nu$. We will not attempt to analyse the transitional regime of the turbulent motion for two reasons. First, because at the present state of knowledge such an analysis is difficult, and second, because it is of hardly any relevance to the civil engineering practice. Indeed, the bed of an estuary or a beach is invariably composed of a loose granular material. If the grain size is large, then the development of the bed irregularities (ripples) might be negligible, but then $X_k \sim k_s$ is still large because the grain size is large. If the grain size is small, then ripples are usually developed, and thus X_k is again appreciable because it is of the order of size of the ripples. In practice there is hardly such a plain and smooth bed where oscillatory fluid motion in the vicinity of the bed, due to waves, could be in any other but rough turbulent regime. The above analysis of the viscous (laminar) motion was not carried out in order to use the relationships determined for the design of practical models; it was carried out

(a) to illustrate the application of the presented method for determining the scales on the well-known, and at the same time simplest, case, and what is more important,
(b) to demonstrate that the difference between the scale relationships corresponding to viscous and rough turbulent motions is appreciable.

* e.g. the criterion given by H. Li[12] can be expressed as $(X_v)_{cr} = 800^{-2} = \frac{1}{64 \cdot 10^4}$.

Today even in some advanced treatises (see eqn. (17.9) in Ref. 9) the similarity criteria, and hence the scales for wave motion are usually derived by using the viscous (laminar) flow expressions (by using $\mu(\partial^2 u/\partial y^2)$ as $\partial\tau/\partial y$ in the equation of motion). But do the scales derived in such a way approximate the truth? If one thinks that the answer is affirmative, because the realistic (rough turbulent) conditions can well be approximated by those of a viscous motion, then the reason pointed out in (b) is certainly justified.

EXAMPLE 7.1 (Aberdeen Channel model; HRS Wallingford)*

Typhoon anchorages, where small Chinese junks and fishing craft can be sheltered, are urgently needed around Hong Kong Island. In order to meet this need, and the need for reclaimed land and quay facilities, in the location of Aberdeen Island (Fig. 7.5a) it is proposed to join the main island of Hong Kong to Applichau (Aberdeen Island) by reclaiming part of the Aberdeen Channel and by building breakwaters across the Southern entrance to the channel. To decide which of the arrangements (shown in Fig. 7.5b) is most effective model tests are needed. Determine the scales of the model required.

If the absolute size of the model is given (i.e. if the model scale λ is selected), then the model wave motion is completely determined by the model wave height H'', the model wave period T'' and the roughness of the model surface k_s''. But the scales λ_H, λ_T and λ_{k_s} are themselves the functions of the model scale λ. Thus, strictly speaking, we can select only one scale, say λ, freely. On the other hand, the tests under consideration concern the behaviour of the waves themselves, they do not concern the behaviour of the fluid motion in the vicinity of the bed. Hence, the model roughness k_s'' (and thus λ_{k_s}) which determines the fluid motion at the bed can, up to a certain extent, be selected at random. Selecting the model scale 1/120 and considering that $\lambda_H = \lambda$ and $\lambda_T = \sqrt{\lambda}$, we arrive at the following scales determining the model

$$\lambda_H = \lambda = \frac{1}{120}$$

$$\lambda_T = \sqrt{\frac{1}{120}} = \frac{1}{10{\cdot}95}$$

for example 10 second period waves in the prototype must be reproduced in the model by approximately 1 second period waves. Let us assume that

* Hydraulics Research 1961, Ministry of Technology, H.M. Stationery Office, London, 1962.

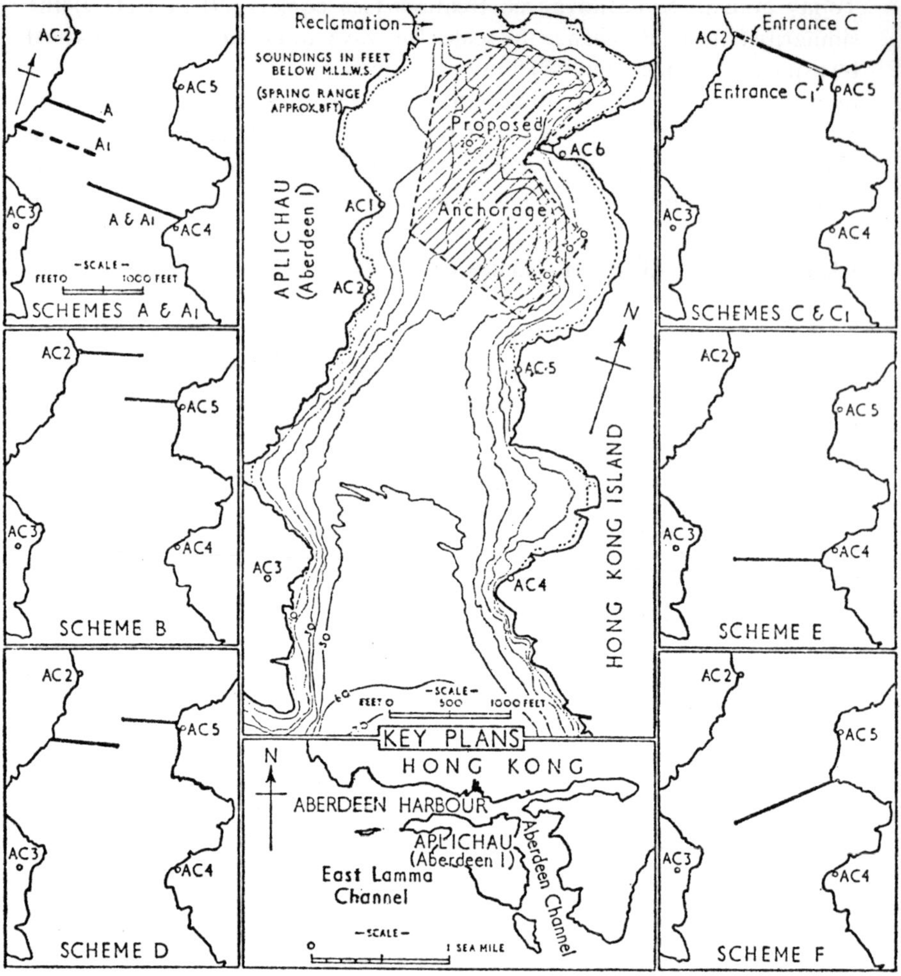

FIG. 7.5 Aberdeen channel, Hong Kong
Breakwater schemes for proposed typhoon shelter

the effective prototype roughness of the harbour bed is of the order of $k_s' = 100$ mm. According to (7.87) the model roughness should be

$$k_s'' = \frac{10}{120} = 0{\cdot}83 \text{ mm}$$

that is, should be slightly less than 1 mm (the presence of rough turbulent motion is assumed). If this requirement is not observed and the roughness

$k_s'' \approx 0{\cdot}5$ mm of the cement mortar model floor is used, then the actual scale of roughness is

$$\lambda_{k_s} = \frac{0{\cdot}05}{10} = \frac{1}{200}$$

(rather than $\lambda_{k_s} = 1/120$). But this merely implies (according to (7.88)) that the scale of the boundary layer thickness is

$$\lambda_\delta = \left(\frac{1}{120}\right)^{\frac{3}{4}}\left(\frac{1}{200}\right)^{\frac{1}{4}} = \frac{1}{36{\cdot}3} \times \frac{1}{3{\cdot}76} = \frac{1}{136{\cdot}4}$$

(rather than $\lambda_\delta = 1/120$ as it should be according to (7.89)). Clearly, the fact that the boundary layer thickness in the model is

$$\frac{136{\cdot}4}{120} = 1{\cdot}14$$

times smaller has no bearing on the testing of surface waves in the (upper) potential flow region. The photograph in (Plate 5) shows the Aberdeen Channel model in operation.

7.4 Long (Tidal) Waves Over a Rigid Bed

It is assumed, in practice, that the period of the long (tidal) wave is sufficiently large for one to assume that δ has reached its ceiling

$$\delta = h_o$$

Accordingly, the concepts of outer and inner regions lose their meaning, and the fluid motion in the whole of the interval

$$(0 \text{ or } k_s) < y < \delta = h_o$$

is defined by the single differential equation

$$\overbrace{\underbrace{u\frac{\partial u}{\partial x}}_{\text{convective acc.}} + \underbrace{\frac{\partial u}{\partial t}}_{\text{temporal acc.}}}^{\text{Total acceleration}} = \overbrace{\underbrace{-\frac{1}{\rho}\frac{\partial p}{\partial x}}_{\text{pressure}} + \underbrace{\frac{1}{\rho}\frac{\partial \tau}{\partial y}}_{\text{friction}}}^{\text{Forces per unit fluid-mass}} \qquad (7.92)$$

In the case of short waves with small amplitude ($\delta \ll h$), the motion in the outer region is given by

$$u\frac{\partial u}{\partial x} + \frac{\partial u}{\partial t} = -\frac{1}{\rho}\frac{\partial p}{\partial x} + 0 \qquad (\delta < y < h)$$

while the motion in the inner region is given by

$$0 + \frac{\partial u}{\partial t} = 0 + \frac{1}{\rho}\frac{\partial \tau}{\partial y} \qquad (0 \text{ or } k_s < y < \delta)$$

which is the eqn. (7.44) considered in the previous section.

Since in the equation (7.92), the vertical convective acceleration is neglected, it is implied automatically that we are dealing with *gradually varying* motion (along x), or with a long wave of *small amplitude* ($r_2 \to 0$), and therefore the instantaneous pressure gradient $-\partial p/\partial x$ is related to the instantaneous slope $-\partial h/\partial x$ of the free surface as follows.

$$-\frac{1}{\rho}\frac{\partial p}{\partial x} = -g\frac{\partial h}{\partial x} \tag{7.93}$$

It is assumed that the bottom is horizontal and thus that the slope $\bar{S}$ of the free surface is equal to $-\partial h/\partial x$.

Substituting (7.93) in (7.92) we arrive at the following version:

$$\overbrace{u\frac{\partial u}{\partial x}}^{\substack{\text{convective}\\ \text{acceleration}}} + \overbrace{\frac{\partial u}{\partial t}}^{\substack{\text{temporal}\\ \text{acceleration}}} = \overbrace{-g\frac{\partial h}{\partial x}}^{\text{pressure}} + \overbrace{\frac{1}{\rho}\frac{\partial \tau}{\partial y}}^{\text{friction}} \tag{7.94}$$

The importance of each term in the equation above is, of course, not the same throughout the thickness of the fluid. When approaching the bed, the streamlines become nearly parallel to the surface of the bed, and thus the importance of the convective term gradually reduces, whereas the importance of the friction term increases.

The eqn. (7.94) can be used for determining the scales more conveniently if it is brought into a dimensionless form. Considering that $\delta = h$ we introduce the following dimensionless variables

$$\xi = \frac{x}{L}, \quad \eta = \frac{y}{h}, \quad \theta = \frac{t}{T} \tag{7.95}$$

and the dimensionless functions

$$\varphi_u = \frac{u}{U} \quad \varphi_h = \frac{h}{h_o} \quad \varphi_\tau = \frac{\tau}{\bar{\tau}_o} \tag{7.96}$$

where U and $\bar{\tau}_o$ are typical velocity and shear stress. Using (7.95) and (7.96) we can express (7.94) as

$$\left[\frac{U^2}{L}\right]\varphi_u\frac{\partial \varphi_u}{\partial \xi} + \left[\frac{U}{T}\right]\frac{\partial \varphi_u}{\partial \theta} = -\left[\frac{gh_o}{L}\right]\frac{\partial \varphi_h}{\partial \xi} + \left[\frac{\bar{\tau}_o}{\rho h_o}\right]\frac{\partial \varphi_\tau}{\partial \eta}$$

* It will be realised that the equations corresponding to the outer and inner regions of the fluid motion due to short waves are two special cases of equation (7.92).

and dividing by the coefficient of $\partial\varphi_u/\partial\theta$ as

$$\overbrace{\left[\frac{UT}{L}\right]\varphi_u\frac{\partial\varphi_u}{\partial\xi}}^{\text{convective acceleration}} + \overbrace{\frac{\partial\varphi_u}{\partial\theta}}^{\text{temporal acceleration}} = -\overbrace{\left[\frac{UT}{L}\right]\left[\frac{U^2}{gh_o}\right]^{-1}\frac{\partial\varphi_h}{\partial\xi}}^{\text{pressure}} + \overbrace{\left[\frac{UT}{h_o}\right]\left[\frac{\bar{\tau}_o}{\rho U^2}\right]\frac{\partial\varphi_\tau}{\partial\eta}}^{\text{friction}} \tag{7.97}$$

It follows that the relationship between the dimensionless depth, velocity and shear stress (between φ_h, φ_u and φ_τ) depends on the numerical values of three dimensionless combinations,

$$\left.\begin{array}{l} \left[\dfrac{U^2}{gh_o}\right] \text{ (Froude number)} \\ \left[\dfrac{UT}{L}\right] \text{ (Strouhal number)} \\ \text{and} \\ \left[\dfrac{UT}{h_o}\right]\left[\dfrac{\tau_o}{\rho U^2}\right] \end{array}\right\} \tag{7.98}$$

rather than on the third combinations alone (with δ instead of h_o) as was the case in Section 7.3.2. It is obvious that the relationship between φ_h, φ_u and φ_τ can be the same in model and prototype only if the numerical values of the dimensionless combinations above are the same. Equating the model and prototype values of these combinations we arrive at

$$\left.\begin{array}{l} \lambda_U = \sqrt{(\lambda_g\lambda_h)} \\ \lambda_T = \sqrt{(\lambda_g\lambda_h)}\dfrac{\lambda_L}{(\lambda_g\lambda_h)} \\ \lambda_\tau = \lambda_\rho(\lambda_g\lambda_h)\dfrac{\lambda_h}{\lambda_L} \end{array}\right\} \tag{7.99}$$

In the case of a practical model we have

$$\lambda_g = \lambda_\rho = 1 \tag{7.100}$$

Furthermore,

$$\lambda_h = \lambda_y \quad \text{while} \quad \lambda_L = \lambda_x \tag{7.101}$$

Substituting (7.100) and (7.101) in (7.99) we obtain

$$\left.\begin{aligned} \lambda_U &= \sqrt{\lambda_y} \\ \lambda_T &= \sqrt{(\lambda_y)}\,\frac{\lambda_x}{\lambda_y} = \sqrt{(\lambda_y)}\,\frac{1}{n} \\ \lambda_\tau &= \lambda_y\,\frac{\lambda_y}{\lambda_x} = \lambda_y n \end{aligned}\right\} \tag{7.102}$$

where n is the distortion.

We assume that the typical values U and $\bar{\tau}_o$ correspond to the same (typical) instant. In this case, if the motion of the fluid is in a rough turbulent regime, then U and $\bar{\tau}_o$ are interrelated through relative roughness. Without restricting the generality of the derivations above we can identify the typical velocity U with the maximum value of the average velocity. By doing so, and by assuming, as is the case in practice, that the motion is in rough turbulent regime, we can express the relation between $U = v_{\max}$ and $\bar{\tau}_o = (\tau_o)_{\max}$ as follows

$$\frac{(\tau_o)_{\max}}{\rho v_{\max}^2} = \frac{1}{\left[2{\cdot}5 \ln\left(11\,\dfrac{h_o}{k_s}\right)\right]^2} = \frac{1}{c^2} \tag{7.103}$$

At this stage it should be pointed out that in the case of long waves, the usual short wave phase shift $T/4$ between the extreme values of the velocity, shear stress and free surface slope diagrams (which represent the variation with time at these quantities at the same section ξ) disappears. In other words, in the case of long waves, maximum values of τ_o, v and $\bar{S}$ occur at the same time (as in unidirectional flows). It has been found[8] that any phase shift becomes practically undetectable when the dimensionless combination LH/h_o^2 exceeds a certain critical value. Clearly, this critical value must itself vary as a function of relative roughness k_s/h_o. In reported experiments[8] the order of the critical value of LH/h_o^2 was found to be ≈ 500. This corresponds to $k_s/h_o = 1/61{\cdot}7$. It is possible that the phase shift between the diagrams of U, $\bar{S}$ and τ_o disappears, when δ reaches its upper limit h_o.

Once again, the logarithmic form can be approximated by the power relationship as

$$c = 2{\cdot}5 \ln\left(11\,\frac{h_o}{k_s}\right) \approx \text{const}\left(\frac{h_o}{k_s}\right)^{\beta} \tag{7.104}$$

Substituting (7.104) in (7.103) we obtain

$$\frac{(\tau_o)_{max}}{\rho v_{max}{}^2} \approx \frac{1}{\left[\text{const}\left(\frac{h_o}{k_s}\right)^{\beta}\right]^2} \qquad (7.106)$$

and thus

$$\lambda_{\tau_o} \approx \lambda_\rho \lambda_U{}^2 \left(\frac{\lambda_{k_s}}{\lambda_h}\right)^{2\beta} \qquad (7.107)$$

Taking into account the values

$$\lambda_\rho = 1 \quad \lambda_h = \lambda_y \quad \lambda_U = \sqrt{\lambda_y}$$

we arrive at the following version of the shear stress scale

$$\lambda_\tau \approx \lambda_{k_s}{}^{2\beta} \lambda_y{}^{1-2\beta} \qquad (7.108)$$

When dealing with short waves we had two versions for λ_τ. One followed from the third combination in (7.98), with δ instead of h_o, the other from the expression (7.59), which is analogous to (7.103), and yet contains δ rather than h_0. Since these two versions had two unknowns λ_δ and λ_τ we have eliminated λ_δ and arrived at only one well defined expression for λ_τ (eqn. (7.82)). In the case of long waves, the consideration of the unknown λ_δ is replaced by that of the known $\lambda_h = \lambda_y$; hence two well defined versions for λ_τ (eqn. (7.108) and the third eqn. in (7.102)).

Equating (7.108) with the third equation in (7.102) we obtain the following value for the roughness scales

$$\lambda_{k_s} \approx \lambda_y n^{1/2\beta}* \qquad (7.109)$$

7.5 Superimposition of Unidirectional Flow with Long and Short Waves (estuary models with rigid bed)

Here, we will assume that each component of the compound flow is in rough turbulent regime. Furthermore we will assume that the small scale model operates with the prototype fluid (water). The following notation is introduced for common quantities (properties and parameters):

Subscript 1 for quantities related to short (wind) waves.
Subscript 2 for quantities related to long (tidal) waves.
Subscript 3 for quantities related to unidirectional flow.

* In future when using (7.109) we will write simply (=) rather than ($\approx$).

The short (wind) waves will always be undistorted; the long (tidal) waves and the unidirectional flow will be distorted (in the same proportion). The scale λ of the undistorted wind waves will be identified with the common vertical scale λ_y of the distorted tidal waves and unidirectional flow.

$$\lambda = \lambda_y \tag{7.110}$$

(i) From the considerations in Section 7.3, it follows that the scales of the parameters specifying the model with the short wind waves ($\delta_1 \ll h_1$) are

$$\left.\begin{aligned} (\lambda_H)_1 &= (\lambda_h)_1 = \lambda_y \\ (\lambda_T)_1 &= \sqrt{\lambda_y} \\ &\text{and} \\ (\lambda_{k_s})_1 & \end{aligned}\right\} \tag{7.111}$$

The scale of any property A must be a certain function of the freely selected scales λ_y and λ_{k_s}. Obviously, the properties related to the outer motion will be the functions of λ_y only. The scales of some relevant properties are

$$\left.\begin{aligned} (\lambda_U)_1 &= \sqrt{\lambda_y} \\ (\lambda_L)_1 &= (\lambda_a)_1 = \lambda_y \\ (\lambda_\delta)_1 &= (\lambda_\tau)_1 = \lambda_y^{\frac{3}{4}}(\lambda_{k_s})_1^{\frac{1}{4}} \\ &\dots\dots\dots\dots \end{aligned}\right\} \tag{7.112}$$

(ii) According to Section 7.4 the scales of the parameters specifying the distorted model with long waves ($\delta_2 = h_2$) are

$$\left.\begin{aligned} (\lambda_H)_2 &= (\lambda_h)_2 = \lambda_y \\ (\lambda_T)_2 &= \sqrt{(\lambda_y)}\,\frac{1}{n} \\ (\lambda_{k_s})_2 &= \lambda_y n^{1/2\beta} \end{aligned}\right\} \tag{7.113}$$

$$(\text{with } n = \lambda_y/\lambda_x)$$

the scales of the properties being certain functions of the freely selected λ_y and n, e.g.

$$\left.\begin{aligned} (\lambda_U)_2 &= \sqrt{\lambda_y} \\ (\lambda_L)_2 &= (\lambda_a)_2 = \lambda_x = \frac{\lambda_y}{n} \\ (\lambda_\tau)_2 &= \lambda_y n \\ &\dots\dots\dots\dots \end{aligned}\right\} \tag{7.114}$$

(iii) From Chapter 5 it is clear that the distorted model of non-uniform and non-stationary unidirectional flow is given by (5.32), or by its generalized version (5.61), and by the Strouhal number. In other words it is determined by

$$\left.\begin{aligned} \lambda_S = \frac{1}{(\lambda_c{}^2)_3} &= n \\ (\lambda_U)_3 &= \sqrt{\lambda_y} \\ (\lambda_t)_3 &= \frac{\lambda_x}{(\lambda_U)_3} \end{aligned}\right\} \tag{7.115}$$

Observe that the third (Strouhal) relationship above, which can be expressed as

$$(\lambda_t)_3 = \frac{\lambda_x}{\sqrt{\lambda_y}} = \sqrt{(\lambda_y)}\,\frac{1}{n} \tag{7.116}$$

coincides precisely with the period scale $(\lambda_T)_2$ of the tidal motion; the scale $(\lambda_U)_3$ being identical to $(\lambda_U)_2$ (and, in fact, to $(\lambda_U)_1$). Hence, the compatibility of (iii) with (ii) depends on whether the first equation of (7.115) is in agreement with the tidal motion or not. Observe that the scale $(\lambda_{k_s})_2$ of the tidal motion follows from

$$(\lambda_\tau)_2 = \lambda_y n$$

and from (7.103), which, according to the present notation, should be expressed as

$$\frac{((\tau_o)_{\max})_2}{\rho(v_{\max}{}^2)_2} = \frac{(\tau_o)_2}{\rho(v^2)_2} = \frac{1}{\left[2{\cdot}5 \ln\left(11\,\dfrac{h_o}{k_s}\right)_2\right]^2} \tag{7.117}$$

The right hand side of the last equation is the reciprocal of the square of the friction factor of the unidirectional flow, and thus it can always be denoted as

$$\frac{(\tau_o)_2}{\rho(v^2)_2} = \frac{1}{(c^2)_3} \tag{7.118}$$

Thus

$$\frac{(\lambda_\tau)_2}{(\lambda_U{}^2)_2} = \frac{1}{(\lambda_c{}^2)_3} \tag{7.119}$$

Substituting here, instead of $(\lambda_T)_2$ and $(\lambda_U)_2$ their values $\lambda_y n$ and $\sqrt{\lambda_y}$, respectively, we arrive at the first equation of (7.115). Hence, all three relations defining the distorted model of unidirectional flow are compatible with those of the tidal flow, and so both these flows may be reproduced simultaneously.

If a flow is gradually varying, then the energy losses are entirely due to friction. Since, in the differential equation (7.92) the energy loss is reflected only by the friction term

$$\frac{1}{\rho}\frac{\partial \tau}{\partial y}$$

we have no alternative but to regard this equation as valid for gradually varying flows only. Observe, however, that in principle the equation (7.94) differs from (1.92) and thus from (5.26) only because of the additional term $\partial u/\partial t$. Thus, eqn. (7.94) can also be generalized into a relation that can represent the fluid motion, even if the geometry of the channel does not vary gradually. Such a generalization can be achieved by introducing additional terms reflecting the local energy losses (i.e. in a manner entirely analogous to the generalization of (5.26) to (5.56)). Hence, from the fact that the form (7.94) is valid for gradually varying flows, should not be inferred that the scale relations above may be utilized only for gradually varying channels (estuaries). On the contrary, the more intensive the variation in the channel geometry, the less sensitive will be the model so far as errors in its roughness are concerned (see Subsection 5.3.4).

Let us now attempt to superimpose the short (wind) waves on the tidal waves and unidirectional flow. From the comparison of (7.111) with (7.113) and (7.115) it is clear that $(\lambda_T)_1$ is not the same as $(\lambda_T)_2$ and $(\lambda_T)_3$. However, we have no alternative but to maintain $(\lambda_T)_1$ as it stands, for only in this case will the velocities due to the short waves be scaled down in the same (Froudian) proportion, $\sqrt{\lambda_y}$, as those due to long waves and unidirectional flow.

Since the model must have a certain (common) roughness, we can have only one roughness scale and thus

$$\lambda_{k_s} = (\lambda_{k_s})_1 = (\lambda_{k_s})_2 = (\lambda_{k_s})_3 = \lambda_y n^{1/2\beta} \qquad (7.120)^*$$

must be valid. Considering these factors we arrive at the following set of scales determining the rigid bed model of the complex

[(short waves) + (tidal waves) + (unidirectional flow)]

* Henceforward scales that are necessarily the same for all three components of the compound motion will not be marked by any subscript at all.

$$\left.\begin{aligned} \lambda_h &= \lambda_y \\ (\lambda_H)_1 &= (\lambda_H)_2 = \lambda_y \\ (\lambda_T)_1 &= \sqrt{\lambda_y} \\ (\lambda_T)_2 &= (\lambda_t)_3 = \sqrt{(\lambda_y)}\,\frac{1}{n} \\ \lambda_{k_s} &= \lambda_y n^{1/2\beta} \end{aligned}\right\} \tag{7.121}$$

The model specified by (7.121) possesses the following properties:

(a) The velocities of all three components of complex motion are scaled down in the same (Froudian) proportions, i.e.

$$(\lambda_U)_1 = (\lambda_U)_2 = (\lambda_U)_3 = \sqrt{\lambda_y} \tag{7.122}$$

is valid.

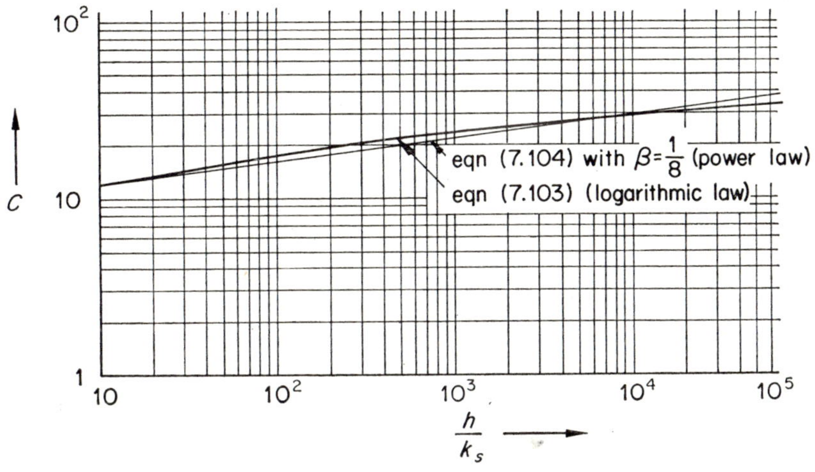

FIG. 7.6

(b) The time scales are related as follows:

$$\frac{(\lambda_T)_1}{(\lambda_T)_2} = \frac{(\lambda_T)_1}{(\lambda_T)_3} = n \tag{7.123}$$

which implies that

$$\frac{1}{\lambda_N} = \frac{\text{(the number of short waves in one tidal period in the prototype)}}{\text{(the number of short waves in one tidal period in the model)}} =$$

$$= \text{distortion}$$

(c) The scale of the boundary layer thickness, and of the shear stresses due to short waves is given by the following relationship (determined by substituting the value of the roughness scale given by (7.121) into the expression of $(\lambda_\delta)_1$ in (7.112))

$$(\lambda_\tau)_1 = (\lambda_\delta)_1 = \lambda_y n^{1/8\beta} \tag{7.124}$$

Observe that this scale of the shear stresses due to short waves can be regarded as being equal to

$$(\lambda_\tau)_2 = (\lambda_\tau)_3 = \lambda_y n$$

if

$$\beta = \tfrac{1}{8} \tag{7.125}$$

From the graph in Fig 7.6 it is clear that the value $\beta = \frac{1}{8}$ approximates the logarithmic function (as implied by (7.104)) in a reasonable manner throughout the range

$$\approx 10 < \frac{h}{k_s} < \approx 10^5 \tag{7.126}$$

Considering that the range above covers virtually all values of h/k_s that can be encountered in the practice, one can assume that the shear stresses are also scaled down in the same proportion

$$(\lambda_\tau)_1 = (\lambda_\tau)_2 = (\lambda_\tau)_3 = \lambda_y n \tag{7.127}$$

EXAMPLE 7.2 ('Telok Anson' model; HRS Wallingford)*

The town, Telok Anson (Malaysia), lies on the spur of land at the river Perak, as indicated in the Key plan in Fig. 7.7. The continuous erosion at the North and South erosion zones threatens to cut the town in two if it is not halted. It is intended to reduce the currents causing erosion, and thus eliminate the danger, by introducing necessary structures (spurs, guide vanes, etc.) and by undertaking the necessary engineering works (dredging, bank protection, etc.). The most effective solution has to be revealed by model tests. Determine the model.

The investigation concerns, primarily the *flow pattern* of the unidirectional and tidal flow in the river. Accordingly, the model can be with a rigid bed, and thus it can be determined by the scale relations (7.121). Two scales, say λ_y and λ_x, can be selected freely. We choose

$$\lambda_y = \frac{1}{60} \quad \text{and} \quad \lambda_x = \frac{1}{600} \qquad (\text{i.e. } n = 10)$$

* Hydraulics Research 1966, Ministry of Technology, H.M. Stationery Office, London, 1967.

Substituting these values in the relations (7.121) we obtain the following values for the rest of the scales determining the model

$$\lambda_h = \frac{1}{60}$$

$$(\lambda_H)_2 = \frac{1}{60}$$

$$(\lambda_T)_2 = (\lambda_t)_3 = \sqrt{\left(\frac{1}{60}\right)}\,\frac{1}{10} = \frac{1}{77{\cdot}5}$$

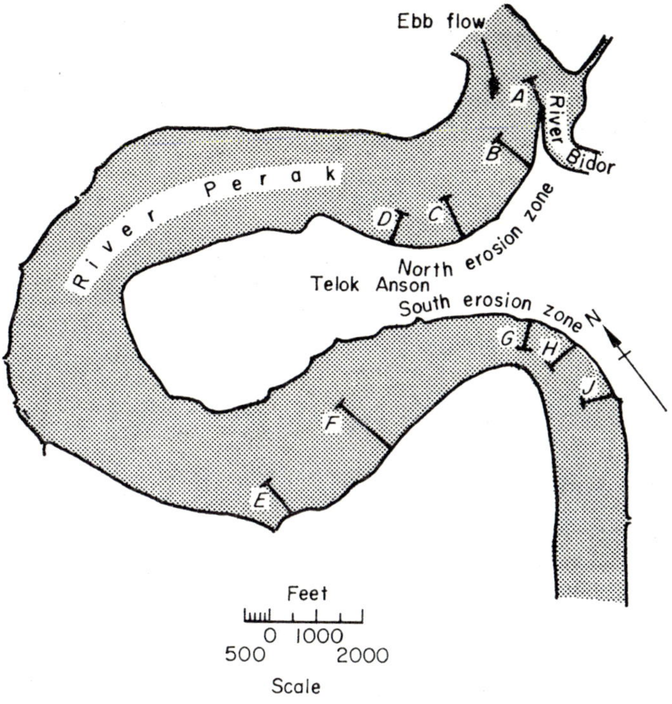

FIG. 7.7 Erosion at Telok Anson, Malaysia
Key plan

Short waves are not present. Hence, the value of β need not necessarily be equal to $\approx 1/8$ in order to achieve compatibility with respect to friction. In addition, the geometry of the river varies considerably along its length, and therefore one can hardly postulate that the energy losses are entirely due to friction. Accordingly, the λ_{k_s} relation in (7.121) does not necessarily need to be fulfilled. The model roughness was therefore determined by means of the experimental trial and error method described in 5.3.3

(Example 5.2). A photograph of the model is shown in Plate 6, its velocity scale being

$$(\lambda_U)_2 = (\lambda_U)_3 = \sqrt{\frac{1}{60}} = \frac{1}{7{\cdot}75}$$

Observe that if waves were present, flow was gradually convergent, and the order of β was $\approx 1/8$, then the distortion $n = 10$ would yield such a large λ_{k_s} and thus k_s'' (see the last relation in (7.121)) that it could not possibly be used. In such cases, one would be compelled to choose n much smaller (e.g. of the order of 1.5 to 2.5 depending on the selected λ_y).

EXAMPLE 7.3 ('Napier Harbour' model; HRS Wallingford)*

Napier Harbour is situated at the Southern end of Hawke Bay in the North Island of New Zealand (Fig. 7.8). The ships moored in this continually

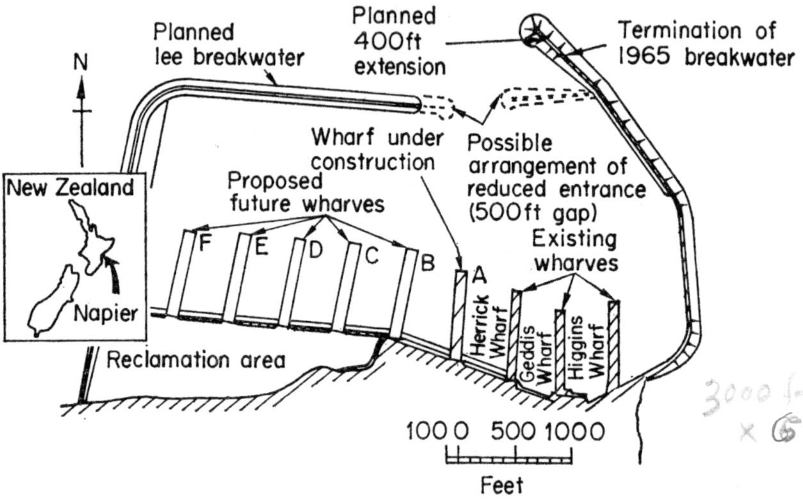

FIG. 7.8 Napier Harbour, New Zealand
Proposed development of harbour

developing port are frequently disturbed by storm waves and long waves, and are occasionally obliged to put to sea to prevent their being damaged. By model tests, it is proposed to determine the most effective arrangement of the breakwaters for the protection of the habour.

For technical reasons the actual tests were carried out in HRS using two different models. One for short storm waves, and another for long waves

* Hydraulics Research 1964, Ministry of Technology, H.M. Stationery Office London, 1965.

(having prototype period of around 2 min). Let us, however, assume that a single model is to be designed for the tests with both long and short waves, and their combination, and let us attempt to determine the scales for such a model. We assume that the waves of 2 min period are sufficiently long in comparison to the average depth of the harbour, and thus that they can be treated in the same way mathematically as the tidal waves. The investigation concerns the influence of the harbour geometry on the behaviour of the waves, and thus of the ships in the harbour. It follows that the use of a rigid bed model should be sufficient.

Two scales can be selected freely and we adopt the scales

$$\lambda_y = \frac{1}{120} \quad \text{and} \quad \lambda_x = \frac{1}{240}$$

corresponding to the (smaller) model in the HRS used for tests with the short waves, (Plate 7). We again utilize the equation (7.121); the difference from the previous example being that we are now dealing with the subscripts 1 and 2 (and not with 2 and 3). The distortion is

$$n = 2$$

and thus the system (7.121) gives

$$\lambda_h = \frac{1}{120}$$

$$(\lambda_H)_1 = (\lambda_H)_2 = \frac{1}{120}$$

$$(\lambda_T)_1 = \sqrt{\left(\frac{1}{120}\right)} = \frac{1}{10{\cdot}96}$$

$$(\lambda_T)_2 = \sqrt{\left(\frac{1}{120}\right)}\frac{1}{2} = \frac{1}{21{\cdot}9}$$

$$\lambda_{k_s} = \frac{1}{120} \times 2^4 \approx \frac{1}{7{\cdot}5}$$

$$(\text{with } \beta = 1/8)$$

It follows that the period of the long waves in the model must be only $T_2'' = 2{\cdot}60/21{\cdot}9 \approx 5{\cdot}5$s, while the model roughness must be 7·5 times smaller than the prototype roughness. This must be so if one needs to achieve similarity for the shear stresses acting on the bed. The tests discussed here however concern the behaviour of the free surface waves themselves, not the action of the fluid on the bed due to these waves. Hence, the condition $\lambda_{k_s} \approx 1/7{\cdot}5$ can be ignored and the tests can be carried out simply by using the cement mortar bed of the model.

Another approach would be to have undistorted long waves. In this case, their model period would be $T_2'' = 11$s and the number of short waves in a long wave would be correct. However, this advantage can hardly compensate the disadvantage of having the scale of the long wave length equal to λ_y when the dimensions of the plan of the harbour (which are comparable with the long wave length) are scaled down in the proportion λ_x.

7.6 Short (Wind) Waves Over a Mobile Bed

Consider the oscillatory two phase motion in the vicinity of the bed generated by short (wind) waves. It is assumed that the geometry of the cohesionless granular material and the shape of the waves is specified. In addition, we assume that δ is much smaller than h_o ($\delta \ll h_o$). In this case the two phase motion under consideration can be completely described by the following set of characteristic parameters

$$\mu,\ \rho,\ D,\ \rho_s,\ \gamma_s,\ a_\delta,\ U_\delta \tag{7.128}$$

This set differs from (7.41) because the consideration of the period T is substituted by the average orbital velocity U_δ, and because the rigid bed parameter k_s is replaced by the mobile bed characteristics D, ρ_s and γ_s.* Selecting ρ, U_δ and D as basic quantities we arrive at the set of four dimensionless variables

$$\left.\begin{aligned} X_1 &= \frac{U_\delta D}{\nu} \\ X_2 &= \frac{\rho U_\delta^2}{\gamma_s D} \\ X_3 &= \frac{a_\delta}{D} \\ X_4 &= \frac{\rho_s}{\rho} \end{aligned}\right\} \tag{7.129}$$

(where $\nu = \mu/\rho$)

which is completely analogous to the set (6.6) for unidirectional flow. The difference is merely due to the fact that instead of the velocity v_* we have the velocity U_δ, while instead of the length h we now have the length a_δ. Here again, the specific mass ρ_s can contribute only to

* As in Chapter 6 the symbol D refers to the grain size in general. The absolute size of the granular material will be specified by giving the value of D_{50}.

the description of the (non-uniform) motion of an isolated grain, and thus the ratio X_4 can no longer be a variable so far as the *mass-motion* of the bed material is concerned. Indeed, from the experimental results[13], it is clear that the similarity of the properties (ripple length Λ, ripple height Δ and net transport rate p) related to the mass motion of the bed material can perfectly well be given by the model and prototype identity of the variables X_1, X_2 and X_3 alone. However, as with unidirectional flow, here too the exclusion of only X_4 is not sufficient for the realization of a small scale model that can be used for practical purposes. Indeed, considering that for a small scale practical model

$$\lambda_v = 1, \quad \lambda_{U_\delta} < 1, \quad \lambda_{a_\delta} < 1 \tag{7.130}$$

must be valid

while
$$\left.\begin{array}{l}\text{the condition } \lambda_{X_1} = 1 \text{ gives } \lambda_D > 1 \\ \text{the condition } \lambda_{X_3} = 1 \text{ gives } \lambda_D < 1\end{array}\right\} \tag{7.131}$$

Thus the realization of a small scale practical model can be achieved only if one of the variables X_1 and X_3 is excluded from consideration. This implies that a small scale model can be designed for the following cases only:

(i) for $\lambda_{X_1} = \lambda_{X_2} = 1$ (if D_{50}' is small and thus μ is relevant) (7.132)

(ii) for $\lambda_{X_3} = \lambda_{X_2} = 1$ (if D_{50}' is large and thus μ is irrelevant) (7.133)

In fact, in case (ii) the condition $\lambda_{X_4} = 1$ is automatically satisfied, and thus in the dynamically similar model implied by (ii) even the properties of the individual grain motion are reproduced without distortion. However, in this and the next three sections, we will deal exclusively with the more difficult case (i), leaving the consideration of (ii) to the end of the chapter (Section 7.9).

Let v_* be the value of the shear velocity corresponding to a specified instant $\theta = t/T$. The quantity v_* is not a property of the motion of an individual grain, and thus is given by

$$\frac{v_*}{U_\delta} = \varphi_*(X_1, X_2, X_3) \tag{7.134}$$

If the model and prototype values of all three variables on the right were identical, as they were in the experiments[13], then the ratio on the left would also be identical, which means that

$$\frac{\lambda v_*}{\lambda_{U_\delta}} = 1$$

i.e.

$$\lambda_{v_*} = \lambda_{U_\delta} \tag{7.135}$$

would be valid. In this case, it would be immaterial which out of the two velocities v_* and U_δ is selected as the characteristic parameter. Indeed, since their scales are equal the conditions (7.132) would always yield the same values for λ_D and λ_{γ_s}, determining the model bed material. However, in contrast to ideal experimental conditions[13], in the practical model under consideration we have the equality of two dimensionless variables only, which implies that the ratio on the left hand side of (7.134) is not identical in model and prototype ($\lambda_{v_*} \neq \lambda_{U\delta}$).

The fact that the scales λ_{v_*} and $\lambda_{U\delta}$ cannot be equal follows also from the earlier considerations. Indeed, in Section 7.3 we have determined

$$\lambda_{U\delta} = \sqrt{\lambda} \quad \text{and} \quad \lambda_{v_*} = \sqrt{\lambda_\tau} = \sqrt{\lambda_\delta}$$

so that $\lambda_{U\delta}$ and λ_{v_*} could be equal only if $\lambda_\delta = \lambda$, i.e. if we have a single length scale in the vicinity of the bed. But this cannot be so unless $\lambda_{a\delta} = \lambda_D$, i.e. $\lambda_{X_3} = 1$.

The question arises: 'Since, in the case of model and prototype, identity of two dimensionless variables (X_1 and X_2) the choice of the characteristic velocity affects the scales λ_D and λ_{γ_s}, which velocity should be used in order to express X_1 and X_2?' The answer to this question can be given from the physical interpretation of the variables X_1 and X_2. Let U be the characteristic velocity we wish to determine. Using U for the expressions of X_1 and X_2 we obtain

$$X_1 = \frac{UD}{\nu} \quad \text{and} \quad X_2 = \frac{\rho U^2}{\gamma_s D} \tag{7.136}$$

Consider now the hydrodynamic force acting on a grain resting on the surface of the bed at any given instant θ. The magnitude of this force is given by

$$F = f\left(\frac{UD}{\nu}\right) \rho D^2 U^2 \tag{7.137}$$

while the weight of the grain in fluid is

$$G = \alpha_f \gamma_s D^3 \tag{7.138}$$

Dividing F by G we obtain

$$\frac{F}{G} = \bar{f}\left(\frac{UD}{\nu}\right) \frac{\rho U^2}{\gamma_s D} \tag{7.139}$$

$$\left(\text{where } \bar{f} = \frac{1}{\alpha_f} f\right)$$

i.e.

$$\frac{F}{G} = \bar{f}(X_1)X_2 \tag{7.140}$$

Thus X_2 can be interpreted as the dimensionless hydrodynamic force exerted by the moving fluid on the bed grain (without drag coefficient), the variable X_1 being the Reynolds number determining the drag coefficient. The projection of hydrodynamic forces (acting on the grains forming the bed surface) on the unit area of the bed surface is simply the shear stress τ_o acting on the surface of the bed. Thus

$$F \sim \tau_o \tag{7.141}$$

or

$$\frac{F}{G_1} \sim \frac{\tau_o}{\gamma_s D} = \frac{\rho v_*^2}{\gamma_s D} \tag{7.142}$$

From comparison of (7.142) and (7.139) it follows clearly that U should be identified with $v_* = \sqrt{(\tau_o/\rho)}$ (as is the case of unidirectional flows). In fact unidirectional flow near the bed can be regarded as a special case of oscillatory motion (corresponding to $T \to \infty$). On the other hand, the oscillatory motion at a time interval $\delta t \to 0$, selected in the vicinity of the instant corresponding to zero acceleration, *is* a unidirectional flow. Considering that in the limiting cases mentioned the expressions of both motions should be compatible, it becomes clear that one should be able to evaluate U, as in the case of a unidirectional flow. In accordance with the explanations above, when referring to X_1 and X_2 henceforth we will always assume that they are expressed by the shear velocity v_* (corresponding to any specified, typical instant)

$$X_1 = \frac{v_* D}{\nu}, \quad X_2 = \frac{\rho v_*^2}{\gamma_s D} \tag{7.143}$$

Let us now consider the effective roughness k_s of the mobile bed. This roughness is also the result of the mass motion of granular material, and thus it is also a function of three variables X_1, X_2 and X_3

$$\frac{k_s}{D} = \varphi_{k_s}(X_1, X_2, X_3) \tag{7.144}$$

(where X_1 and X_2 are expressed by v_*). Since X_1 and X_2 are identical in model and prototype, the multiplier m in the scale relationship

$$\lambda_{k_s} = m\lambda_D \tag{7.145}$$

given by (7.144) will be dependent only on the difference of the model and prototype values of X_3. The length Λ of the ripples due to the short waves always increases when X_3 increases (Fig. 7.9a). On the other hand, the ripple height Δ first increases, then decreases (Fig. 7.9b). Hence, the roughness k_s/D must first increase and then decrease (in a manner shown

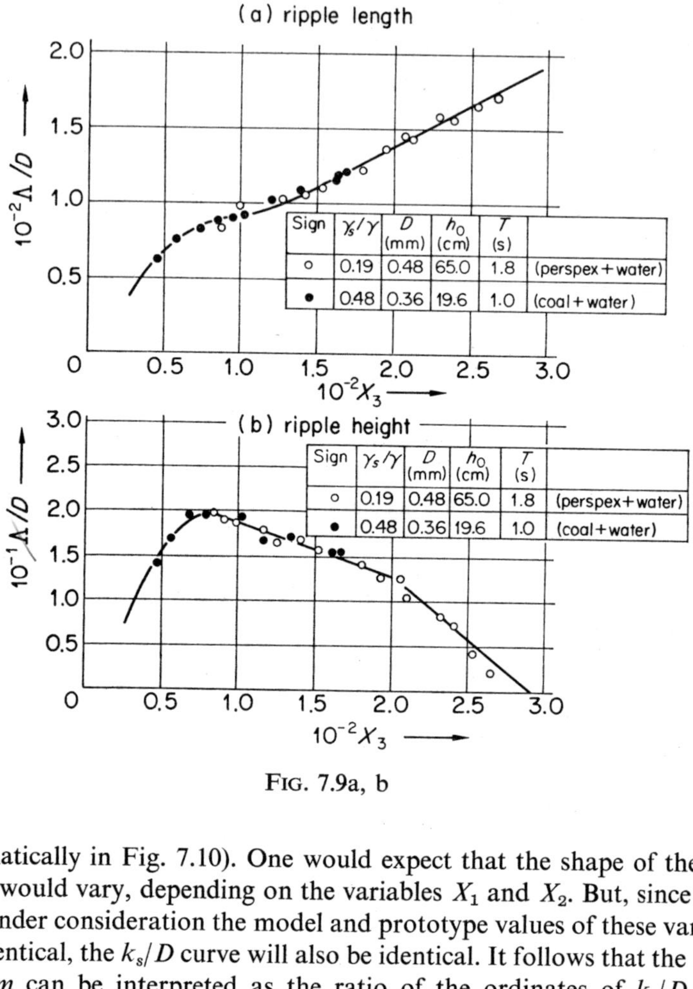

FIG. 7.9a, b

schematically in Fig. 7.10). One would expect that the shape of the k_s/D curve would vary, depending on the variables X_1 and X_2. But, since in the case under consideration the model and prototype values of these variables are identical, the k_s/D curve will also be identical. It follows that the multiplier m can be interpreted as the ratio of the ordinates of k_s/D curve corresponding to the different abscissa:

$$m = \frac{\overline{A''B''}}{\overline{A'B'}} \tag{7.146}$$

Since for a small scale model designed according to (7.132)

$$\lambda_D > 1 \quad \text{and} \quad \lambda_{a_\delta} < 1$$

we have, necessarily

$$\lambda_{X_3} < 1 \quad \text{or} \quad X_3'' < X_3' \tag{7.147}$$

as indicated in Fig. 7.10. However, since the k_s/D curve first rises and then falls, the value of m can be both larger and smaller than unity.

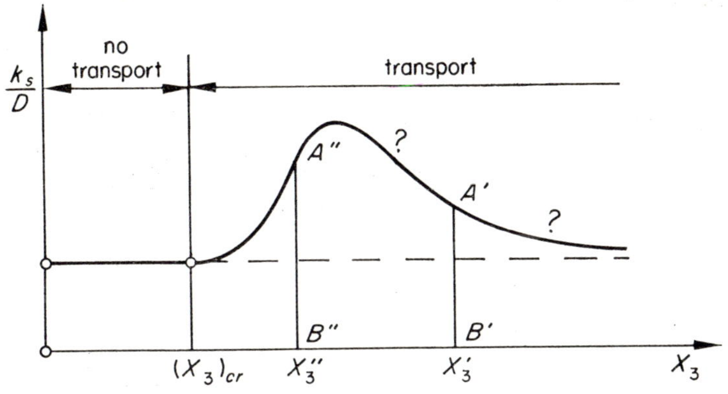

FIG. 7.10

In the present state of knowledge, it is not known how the exact shape of the k_s/D curve should be determined (for the given orders of X_1 and X_2). However, even if such knowledge were available it could hardly help much so far as the improvement of accuracy in model tests is concerned. Indeed, in the case of short (wind) waves the amplitude a_δ (and thus X_3) varies with the depth appreciably. On the other hand, the bed of a natural beach or estuary is far from being horizontal. Thus it is almost certain that in the region under investigation one will encounter a whole spectrum of the values $X_3 = a_\delta/D$, and consequently a whole spectrum of all possible stages of the development of ripples (and thus of the ratio m) as implied by Fig. 7.11. But the model can be designed according to only one value of m (for we can have only one scale $\lambda_{k_s} = m\lambda_D$). Hence, even if our knowledge was far more advanced, we would still be compelled to design the model according to a single (average) value of m, and thus we would still be able to achieve a satisfactory reproduction only for a particular location (having m that coincides with the selected value for m).

In the case of short waves the value of the shear stress scale is given by (7.82) as follows:

$$\lambda_\tau = {\lambda_{v_*}}^2 = \lambda^{\frac{3}{4}} {\lambda_{k_s}}^{\frac{1}{4}}$$

Substituting here the value of λ_{k_s} given by (7.145) we obtain

$$\lambda_\tau = {\lambda_{v_*}}^2 = m^{\frac{1}{4}} \lambda^{\frac{3}{4}} {\lambda_D}^{\frac{1}{4}} \tag{7.148}$$

Fortunately, the exponent of m is relatively small and thus the error is considerably reduced.

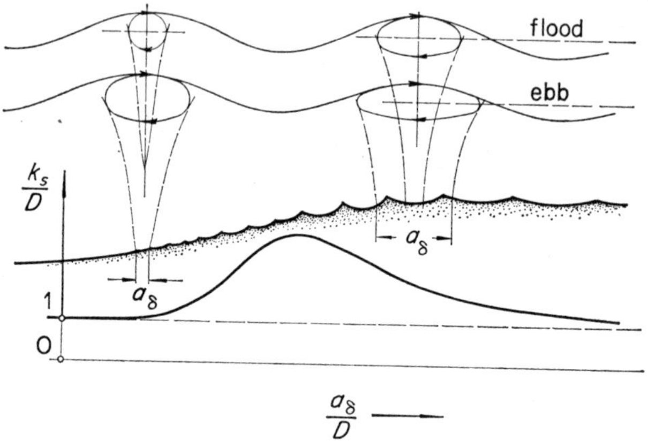

FIG. 7.11

The equality of X_1 and X_2 in model and prototype (formed by v_*), i.e. the conditions (7.132), give

$$\lambda_D = \frac{1}{\lambda_{v_*}} \quad \text{and} \quad \lambda_{\gamma_s} = {\lambda_{v_*}}^3 \tag{7.149}$$

Solving λ_D, λ_{γ_s} and λ_τ from three eqns. (7.148) and (7.149) we obtain the following values for these scales (in terms of the model scale λ and m)

$$\left.\begin{aligned} \lambda_\tau &= m^{\frac{2}{9}} \lambda^{\frac{2}{3}} \\ \lambda_D &= \frac{1}{m^{\frac{1}{9}} \lambda^{\frac{1}{3}}} \\ \lambda_{\gamma_s} &= m^{\frac{1}{3}} \lambda \end{aligned}\right\} \tag{7.150}$$

7.7 Long (Tidal) Waves Over a Mobile Bed

If the geometry of the channel (estuary), of the wave and of the granular material is specified, then two-phase motion due to the tidal wave is completely determined by the following eight characteristic parameters

$$H, T, h_o, \rho, \mu, \rho_s, \gamma_s, D \tag{7.151}$$

Since the consideration of any of these parameters can be substituted by any of their functions, we can, for example, substitute H by a typical (say maximum) value of v_*. In this case we will arrive at

$$v_*, T, h_o, \rho, \mu, \rho_s, \gamma_s, D \tag{7.152}$$

and thus at

$$\left.\begin{aligned} X_1 &= \frac{v_* D}{\nu} \\ X_2 &= \frac{\rho v_*^2}{\gamma_s D} \\ X_3 &= \frac{h_o}{D} \\ X_4 &= \frac{\rho_s}{\rho} \\ X_5 &= \frac{v_* T}{h_o} \end{aligned}\right\} \tag{7.153}$$

But the set of the dimensionless variables above is identical to that of a *non-steady* unidirectional two phase motion. Indeed, the first four dimensionless variables are precisely those studied in Chapter 6, while the fifth is simply the Strouhal number, which serves only in order to bring the typical time into a dimensionless form. From this identity of the scales, one must not infer, however, that the structure of tidal flow is the same as unidirectional flow. On the contrary their structure is different and their corresponding properties are given by different mathematical relations. However, these relations are different functions of the same dimensionless variables (recall Fig. 1.5 showing the variety of flows which correspond to the different origins and structures and yet to the same dimensionless forms). From the compatibility of tidal and unidirectional flows shown in Section 7.5 in rigid bed models, and from the identity of the dimensionless variables mentioned above, it follows clearly that the flows in question must also be compatible in the case of a mobile bed. We will not pursue this matter by further examples and will conclude with a brief reference to the scale of roughness.

The dimensionless roughness due to the tidal flow can be expressed as follows:

$$\frac{k_s}{D} = \bar{\varphi}_{k_s}(X_1, X_2, X_3, \xi, \theta) \tag{7.154}$$

where ξ and θ are the dimensionless location and instant. Let us assign to the function above, the following form

$$\frac{k_s}{D} = X_3^{m_1}\varphi_{k_s}(X_1, X_2, \xi, \theta) \tag{7.155}$$

The tidal period T, and thus X_5, are very large; it is therefore unlikely that acceleration can have any noticeable influence on the formation of the bed features and thus of k_s. Hence, X_5 is not included in the function φ_{k_s}.

It is assumed that the exponent m_1 of X_3 may be a function of X_1 and X_2 but not of X_3. The tidal model will be designed so as to have X_1 and X_2 identical in model and prototype. Furthermore, the model and prototype comparison must necessarily be made for corresponding instants and locations. Hence, the values of m_1 and φ_{k_s} are the same in model and prototype, and therefore (7.155) gives the following value for the scale of

$$\frac{\lambda_{k_s}}{\lambda_D} = \lambda_{X_3}{}^{m_1} \tag{7.156}$$

which depends only on the ratio of the non-equal model and prototype values of the dimensionless variable X_3. Using the value $X_3 = h_o/D$ in the relationship above, we obtain

$$\lambda_{k_s} = \lambda_D{}^{1-m_1}\lambda_h{}^{m_1}$$

or

$$\lambda_{k_s} = \lambda_D{}^{\bar{m}}\lambda_h{}^{1-\bar{m}} \tag{7.157}$$

$$(\text{with } \bar{m} = 1 - m_1)$$

From the Section 7.4 we have the following version for λ_{k_s} (eqn. 7.109)

$$\lambda_{k_s} = \lambda_y n^{1/2\beta} \tag{7.158}$$

while the value of λ_D is given by

$$\lambda_D = \frac{1}{\lambda_{v*}} = \frac{1}{\sqrt{(\lambda_y n)}} \tag{7.159}$$

since, in the case of the long waves

$$\lambda_T = \lambda_{v_*}{}^2 = \lambda_y n$$

Equating (7.157) and (7.158), taking into account (7.159) and considering that $\lambda_h = \lambda_y$, we arrive at the following value of distortion

$$\frac{\lambda_y}{\lambda_x} = \frac{1}{\lambda_y^{3\overline{m}\beta/(1+\overline{m}\beta)}} \tag{7.160}$$

7.8 Superimposition of Unidirectional Flow with Long and Short Waves (estuary models with mobile bed)

7.8.1 Scale Relationships We return to the notation used in Section 7.5:

Subscript 1 for short (wind) waves
Subscript 2 for long (tidal) waves

The subscript 3 (for unidirectional flow) will be omitted in the present section on the grounds that the similarity criteria of non-stationary unidirectional flow coincide with those of tidal motion. A common property, having the same value in both components of the compound motion, will not be marked by any subscript. Here again the short waves are undistorted, their scale λ being identical to the vertical model scale λ_y. It is assumed, as in the preceding sections, that the compound motion of the fluid is rough turbulent, and that we are dealing with the small values of D_{50}' (in the sense implied by (7.132)). In models such as those being considered, the flow is rough turbulent and thus independent of μ, yet the behaviour of the granular bed material is dependent on X_1 and thus on μ. There seems to be an incompatibility here, in that the behaviour of a particle in a flow which is independent of a parameter, is dependent on that parameter! The confusion lies in the attempt to find an explanation by means of using a dimensional quantity rather than using the dimensionless combination (reflecting the influence of that quantity). It is not μ itself, but the magnitude of the Reynolds number that matters when considering the influence on μ. Whether the flow is rough turbulent or not depends on the geometry of the flow as a whole and on the Reynolds number

$$\mathrm{Re}_* = \frac{v_* k_s}{\nu}$$

the Reynolds number determining the behaviour of a grain being

$$X_1 = \frac{v_* D}{\nu}$$

If $k_s \gg D$, then $\mathrm{Re}_* \gg X_1$ and the behaviour (detachment motion, etc.) of *grain in the fluid* will be under the influence of viscosity even though the motion of *fluid in the channel* is independent of it. The fact that the motion of the fluid past the bed irregularities (k_s) is turbulent, does not mean that the detachment and the motion of the grains (of much smaller size than k_s) through this turbulent medium must also be independent of μ.

(i) From the content of the Section 7.6 (and 7.3) it is clear that the scales specifying the model with the short waves ($\delta_1 \ll h_1$) over a mobile bed are

$$\left.\begin{aligned} (\lambda_H)_1 &= (\lambda_h)_1 = \lambda_y \\ (\lambda_T)_1 &= \sqrt{\lambda_y} \\ \lambda_D &= \frac{1}{m^{\frac{1}{9}}\lambda_y^{\frac{1}{3}}} \\ \lambda_{\gamma_s} &= m^{\frac{1}{3}}\lambda_y \end{aligned}\right\} \tag{7.161}$$

some relevant properties being

$$\left.\begin{aligned} (\lambda_U)_1 &= \sqrt{\lambda_y} \\ (\lambda_L)_1 &= (\lambda_a)_1 = \lambda_y \\ (\lambda_\delta)_1 &= (\lambda_\tau)_1 = m^{2/9}\lambda_y^{2/3} \\ (\lambda_{k_s})_1 &= \frac{m^{8/9}}{\lambda_y^{1/3}} \quad * \\ &\cdots\cdots\cdots\cdots\cdots\cdots \end{aligned}\right\} \tag{7.162}$$

(ii) According to the Section 7.7 (and 7.4) the scales determining the model with the long waves ($\delta_2 = h$) over a mobile bed are

$$\left.\begin{aligned} (\lambda_H)_2 &= (\lambda_h)_2 = \lambda_y \\ (\lambda_T)_2 &= \sqrt{\lambda_y} \\ \lambda_D &= \frac{1}{\sqrt{(\lambda_y n)}} \\ \lambda_{\gamma_s} &= (\lambda_y n)^{3/2} \end{aligned}\right\} \tag{7.163}$$

where n and λ_y are interrelated by (7.160) i.e. by

$$n = \frac{1}{\lambda_y^{3\bar{m}\beta/(1+\bar{m}\beta)}} \tag{7.164}$$

* Follows from (7.145) and the second eqn. in (7.150).

Some relevant properties being

$$\left.\begin{aligned} (\lambda_U)_2 &= \sqrt{\lambda_y} \\ (\lambda_L)_2 &= (\lambda_a)_2 = \lambda_x = \frac{\lambda_y}{n} \\ (\lambda_T)_2 &= \lambda_y n \\ (\lambda_{k_s})_2 &= \lambda_y n^{1/2\beta} \\ &\ldots\ldots\ldots\ldots\ldots\ldots \end{aligned}\right\} \tag{7.165}$$

From the scale relations above it is clear that the mobile bed model of the complex

$$[\underbrace{\text{(short waves)}}_{\text{subscript 1}} + \underbrace{\text{(tidal waves)} + \text{(unidirectional flow)}}_{\text{subscript 2}}]$$

is determined by the following scale relations

$$\left.\begin{aligned} (\lambda_H)_1 &= (\lambda_H)_2 = \lambda_h = \lambda_y \\ (\lambda_T)_1 &= \sqrt{\lambda_y} \\ (\lambda_T)_2 &= \sqrt{(\lambda_y)}\,\frac{1}{n} \\ \lambda_D &= \frac{1}{\sqrt{(\lambda_y n)}} \\ \lambda_{\gamma_s} &= (\lambda_y n)^{\frac{3}{2}} \end{aligned}\right\} \tag{7.166}$$

provided that the relation (7.164) is valid.

7.8.2 Discussion Let us first consider the relationship (7.164), and for this purpose examine the meaning of the product $\bar{m}\beta$ which appears in this equation. Substituting the value of k_s given by (7.155) into (7.104), and taking into account that $\bar{m} = 1 - m_1$, we obtain

$$(c)_2 = A_2 \left(\frac{h}{D}\right)^{\bar{m}\beta} \tag{7.167*}$$

(where $A_2 = \text{const}/[\varphi_{k_s}]^{\beta}$)

Hence, $\bar{m}\beta$ is an exponent indicating how the friction factor c varies with the ratio h/D. For the given bed material and fluid, the value of $\bar{m}\beta$ varies with the stage of the flow, for it is a function of X_1 and X_2 (corresponding to the undirectional and tidal flows). However, from the experimental data it follows that for a considerable range of h/D, corresponding to fully

* Which is expressed for any h (not necessarily for h_o).

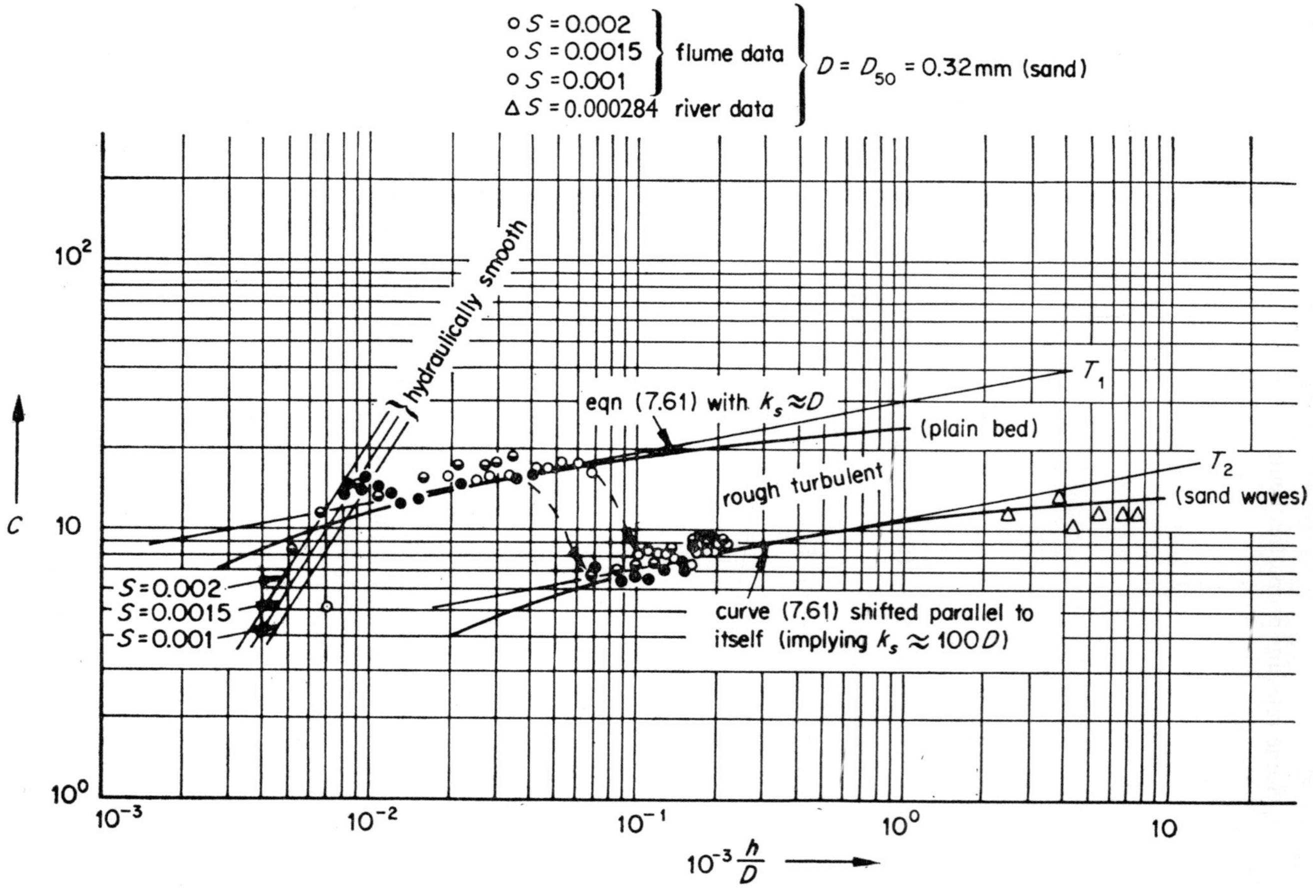

○ S = 0.002
○ S = 0.0015
○ S = 0.001
flume data
Δ S = 0.000284 river data
D = D50 = 0.32mm (sand)
hydraulically smooth
eqn (7.61) with ks ≈ D
(plain bed)
T1
T2
(sand waves)
rough turbulent
S = 0.002
S = 0.0015
S = 0.001
curve (7.61) shifted parallel to itself (implying ks ≈ 100D)
C
10^2
10
10^0
10^-3
10^-2
10^-1
10^0
10
10^-3 h/D

FIG. 7.12

developed sand waves, the value of $\bar{m}\beta$ does not vary appreciably (especially if the sand waves are *ripples*, and the variation of their shape and size with h can thus be ignored). Usually

$$\bar{m}\beta \approx \tfrac{1}{4} \text{ to } \tfrac{1}{6} \tag{7.168}$$

as may be seen, for example, from Fig. 7.12.* At present we are dealing with small grain sizes and with flows varying their direction. Under these conditions it is very likely that the sand waves are ripples (rather than dunes), and thus the approximate value above can be adopted. However, this should be done only if the prototype measurements cannot be carried out. Indeed, the most reliable approach would be to determine from the prototype measurements the variation of the friction factor with relative depths h/D which will undergo model tests, and from the graph of this variation obtain the magnitude of $\bar{m}\beta$.

Assuming that $\bar{m}\beta$ is equal to 1/4 and 1/6 respectively, we obtain from eqn. (7.164)

$$n = \frac{1}{\lambda_y^{\frac{3}{8}}} \text{ (if } \bar{m}\beta = \tfrac{1}{4}) \quad \text{and} \quad n = \frac{1}{\lambda_y^{\frac{3}{7}}} \text{ (if } \bar{m}\beta = \tfrac{1}{6}) \tag{7.169}$$

and thus

$$\lambda_x = \lambda_y^{1.60} \text{ (if } \bar{m}\beta = \tfrac{1}{4}) \quad \text{and} \quad \lambda_x = \lambda_y^{1.43} \text{ (if } \bar{m}\beta = \tfrac{1}{6})^{**} \tag{7.170}$$

The scales λ_x and λ_y of thirty distorted tidal models operating in HRS-Wallingford during the period 1956–67 are plotted in Fig. 7.13. So far, no generally accepted design method for distorted tidal models has been established, and therefore many of the plotted models were designed by using a different approach. It is remarkable that the points in Fig. 7.13 are fairly evenly scattered about the straight line representing the Lacey equation

$$\lambda_x = \lambda_y^{\frac{3}{2}} \tag{7.171}$$

* Unfortunately Fig. 7.12 contains the ripple data of a unidirectional flow only. This graph was determined from the original of Fig. 4, Ref. 20.

** Observe from Fig. 7.12 that the slope of the tangent T_2 is of the same order ($\frac{1}{4}$ to $\frac{1}{6}$) as that of T_1 which corresponds to the stage when the bed is still plain (before the decrement of the value of c due to the development of sand waves, and thus due to the increment of k_s). Hence, the same value of $\bar{m}\beta$ could have been obtained if one started (in principle, incorrectly) from the consideration of the tangent T_1 corresponding to the plain bed. In fact, using the plain rigid bed formula of Strickler ($c_r = \text{const}/D^{1/6}$), S. Z. H. Rizvi[14] has succeeded in arriving at the relation $\lambda_x = \lambda_y^{10/7}$ which is precisely the same as the second equation in (7.170). For the derivation of the scales λ_D and λ_{γ_s}, Rizvi[14] adopts the method similar to that used by E. W. Bijker[15] and uses H. C. Frijlink's formula[16] for the transport rate. However, since the formula mentioned is but a certain function of X_1 and X_2, the expressions for λ_D and λ_{γ_s} derived by the authors mentioned are identical to those derived in the present text directly from the model and prototype equalities of X_1 and X_2.

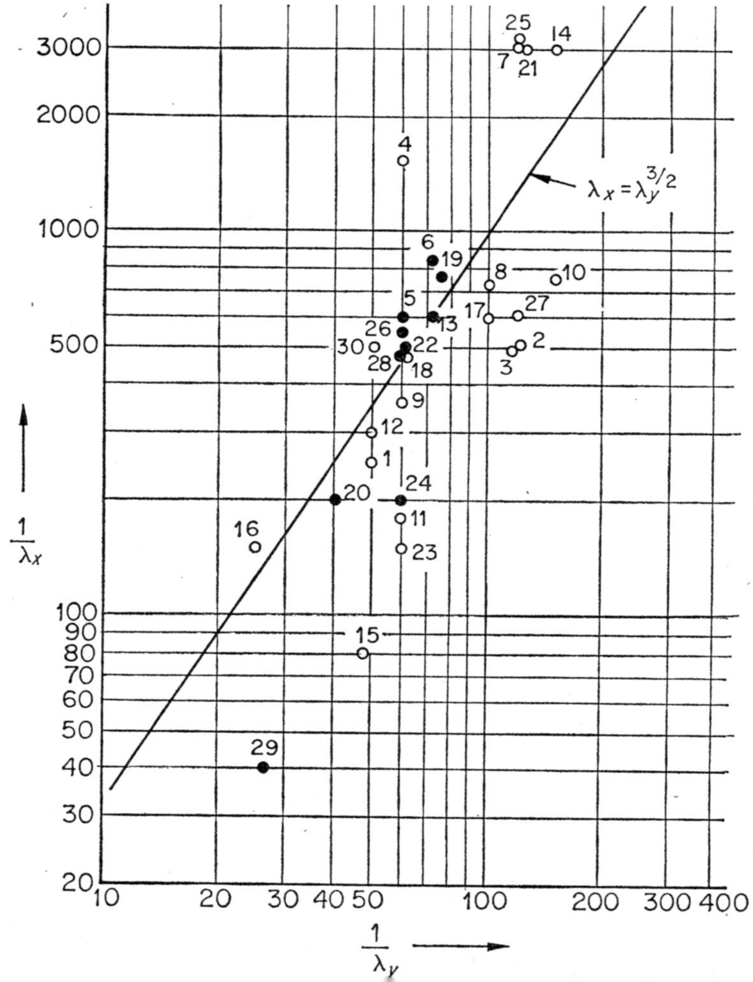

1. Desalination Plant, North Point, Hong Kong (1966)
2. Aberdeen Harbour, Hong Kong
3. Plymouth Sound
4. Crossing of the Dee Estuary
5. Telok Anson, Malaysia (bakelite)
6. Tauranga Harbour, New Zealand (coal)
7. Tauranga Harbour, New Zealand (pilot model)
8. Lower Waikato River, New Zealand
9. Christchurch, New Zealand
10. River Orinoco, Venezuela
11. Plover Cove, Hong Kong
12. Ringsend Power Station, Dublin
13. Portbury, near Avonmouth
14. Tyneside Sea Outfalls
15. River Bann, Northern Ireland
16. Damhead Creek
17. Newcastle Harbour, New South Wales
18. Hong Kong Harbour
19. The Tees Estuary
20. Karachi Harbour
21. Thames Estuary (pilot model)
22. Karnafuli River
23. Conway Harbour
24. The Tees Estuary
25. Liverpool Bay, Mersey Estuary
26. Bromborough Bar, Mersey Estuary
27. Tilbury Power Station, Thames
28. Porthomouth Harbour
29. Eyemouth Harbour
30. Kuala Baram

○ Models with rigid bed

● Models with mobile bed

which corresponds to $1/\bar{m}\beta = 5$. Hence the equation of Lacey can be considered as the 'average value' with regard to the scale relations used in HRS in the past twelve years. Attention is drawn to the fact that the points are not *on* but *around* the straight line representing (7.171). In other words the eqn. (7.171), as it stands, was rarely used, yet it certainly turned out to be the compromise of all the methods used. Apparently the origin of eqn. (7.171) is purely empirical (or even intuitive) and this might be the reason why it has not gained the popularity it deserves. On the other hand, as is clear from the preceding considerations, eqn. (7.171) can be obtained in fact, from the theoretically determined form (7.164) by substituting a single experimental value $\bar{m}\beta = 1/5$.

Consider the equality

$$(\lambda_\tau)_1 = (\lambda_\tau)_2$$

which, taking into account the third equation in (7.162) and (7.165), can be written as

$$\lambda_y n = m^{2/9}\lambda_y^{2/3} \tag{7.172}$$

i.e. as

$$m = (n\lambda_y^{1/3})^{9/2}$$

and by virtue of (7.164) as

$$m = \lambda_y^{3(1-8\bar{m}\beta)/2(1+\bar{m}\beta)} \tag{7.173}$$

If, at the beginning of sediment transport, the bed is plain, then

$$m = 1 \quad \bar{m} = 1 \quad \beta \approx \tfrac{1}{8} \tag{7.174}$$

and thus the conditions (7.173) and (7.172) can be regarded as satisfied. Unfortunately, no such claim can be made if the bed is covered by sand waves. Indeed, in the latter case, the condition (7.172) will be fulfilled only in those regions where the ripples due to short waves are such as to satisfy (7.173).

What will be wrong in the regions where the condition (7.172) is not satisfied? In order to answer this question, let us consider the expression of transport rate. The bed material is lifted (eroded) by the shear stress τ_0, is transported by the flow velocities u. Denoting the typical flow velocity (in the region of the flow occupied by the transported material) by u_s we can write for the transport rate (solids discharge)

$$p \sim \tau_0 u_s \tag{7.175}$$

where the dimensionless proportionality factor is a certain function of the dimensionless variables (but varies most intensively with X_1 and X_2). If the scales of τ_0 and u_s were identical for long and short waves, the scales of the transport rate would also be identical. In the present case, however, when transport is established, it is the identity of the u_s-scale only that is ensured. Hence, with respect to the similarity of the solids discharge p, we can be sure only that the short waves contribute correctly to the transport of the lifted material from one location to another; we cannot be sure that the amount of granular material that they are lifting (in any part of the region under investigation when the bed is no longer plain) is also correct. In the preceding considerations, it has been shown that $(\lambda_\tau)_1$ and $(\lambda_\tau)_2$ can be different. The fact that we have emphasized that among these different shear stresses (and corresponding erosion and sediment transport), the *correct* one corresponds to the tidal and unidirectional flow, is because the model bed material is selected according to $(\lambda_\tau)_2 = \lambda_y n$ (see the values λ_D and λ_{γ_s} in (7.166)), and because of the compatibility implied by the relation (7.164).

The *nett* transport of granular material from one location to another due to short waves is caused by the drift velocity w_D, which is given by the following expression of M. S. Longuet-Higgins[17]

$$w_D = \text{const} \, . \frac{(U_\delta)_1{}^2}{C_1} \tag{7.176}$$

where C_1 is the velocity of propagation of the short waves

$$C_1 = \frac{L_1}{T_1} \tag{7.177}$$

Since the short waves are undistorted, it is obvious that

$$\lambda_{C_1} = (\lambda_U)_1 = \sqrt{\lambda_y} \tag{7.178}$$

is valid, and thus the scale of the drift velocity coincides, indeed, with the scale of the velocities.

From (7.164) it is clear that each particular value of $\bar{m}\beta$ requires its own distortion, i.e. its own power relation between the horizontal and vertical model scales. The straight lines representing various power relations between λ_y and λ_x, corresponding to various values of $\bar{m}\beta$ are shown in a logarithmic system of coordinates in Fig. 7.14. Knowing, from the prototype measurements, the value of $\bar{m}\beta$, one knows the position of the corresponding straight line (the exponent of the power relation between λ_y and λ_x), and one can choose λ_y and λ_x accordingly. If the model has to be designed without knowing the actual prototype value of $\bar{m}\beta$, then it would be advisable to assume the validity of (7.168), and thus to choose λ_y and λ_x from the shaded region in Fig. 7.14. Assuming that the lightest model bed material is polystyrene, we have the condition

$$\lambda_D \leqslant 3{\cdot}8 \tag{7.179}$$

Eq. 7.164 ?

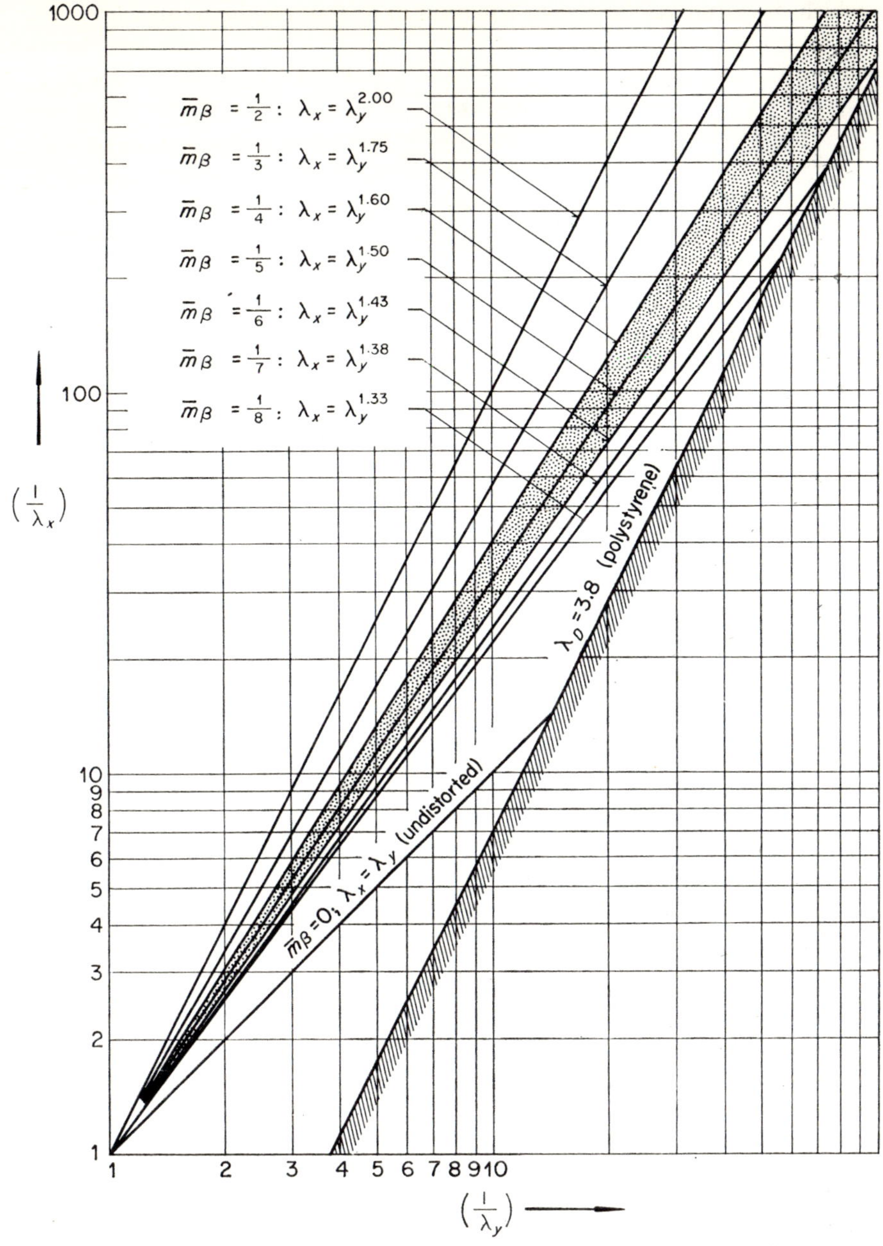

1000
100
10
9
8
7
6
5
4
3
2
1
$\bar{m}\beta = \frac{1}{2}$: $\lambda_x = \lambda_y^{2.00}$
$\bar{m}\beta = \frac{1}{3}$: $\lambda_x = \lambda_y^{1.75}$
$\bar{m}\beta = \frac{1}{4}$: $\lambda_x = \lambda_y^{1.60}$
$\bar{m}\beta = \frac{1}{5}$: $\lambda_x = \lambda_y^{1.50}$
$\bar{m}\beta = \frac{1}{6}$: $\lambda_x = \lambda_y^{1.43}$
$\bar{m}\beta = \frac{1}{7}$: $\lambda_x = \lambda_y^{1.38}$
$\bar{m}\beta = \frac{1}{8}$: $\lambda_x = \lambda_y^{1.33}$
$\left(\frac{1}{\lambda_x}\right)$
$\lambda_0 = 3.8$ (polystyrene)
$\bar{m}\beta = 0$; $\lambda_x = \lambda_y$ (undistorted)
1 2 3 4 5 6 7 8 9 10
$\left(\frac{1}{\lambda_y}\right)$

FIG. 7.14

which, by virtue of the fourth equation in (7.166) can be written as

$$\frac{1}{\sqrt{(\lambda_y n)}} \leqslant 3{\cdot}8 \tag{7.180}$$

i.e. as

$$\frac{1}{\lambda_x} \geqslant \frac{1}{(3{\cdot}8)^2}\left(\frac{1}{\lambda_y}\right)^2 \tag{7.181}$$

As seen from the position of the straight line $\lambda_D = 3{\cdot}8$, in Fig. 7.14, the fulfilment of the condition (7.181) can hardly be regarded as difficult.

For the models studied in this section, the similarity of friction (roughness and shear stresses) has been considered, at least with regard to the tidal motion and unidirectional flow; the order of $\bar{m}\beta$ being important in order to ensure the similarity of friction. Observe that this kind of model can be undistorted only if $\bar{m}\beta = 0$, i.e. if the range of stages under consideration is in the vicinity of the minimum of the c-curve (Fig. 7.12) where

$$\frac{\partial c}{\partial\left(\dfrac{h}{D}\right)} = 0$$

Clearly, this is rather a narrow range to be worthwhile considering while its reproduction (by an undistorted model) requires very big model scales (see Fig. 7.14).

7.8.3 Time Scales Related to the Formation of the Mobile Bed Surface It is intended, first, to establish the time scales related to the formation of the bed surface due to the tidal waves, then to consider them with respect to the combined motion.

(a) *Formation in the direction of y-axis*
If the time scale were $(\lambda_t)_s$ (given by (6.64)), then the scale of erosion and/or accretion would be given by (6.76), that is, by

$$\lambda_Y = \frac{\lambda_D{}^2}{\lambda_x}$$

However, the time scale of the tidal motion is

$$(\lambda_T)_2 = \sqrt{(\lambda_y)}\,\frac{1}{n} \tag{7.182}$$

Thus, the scale of erosion and/or accretion after one tidal period is

$$(\lambda_Y)_2 = \lambda_Y \frac{(\lambda_T)_2}{(\lambda_t)_s} \tag{7.183}$$

and consequently

$$(\lambda_Y)_2 = \frac{\lambda_D{}^2}{\lambda_x} \frac{\sqrt{\lambda_y}}{n} \frac{\lambda_{v_*}}{\lambda_D} = \frac{1}{\sqrt{\lambda_y}} \tag{7.184}$$

It would be desirable to have the erosion and/or accretion scale equal to the vertical model scale λ_y. To make it possible, let us consider a duration $\mathscr{T}$ consisting of N_y tidal periods. We have

$$\mathscr{T} = N_y(T)_2 \tag{7.185}$$

and thus

$$\lambda_{\mathscr{T}} = \lambda_{N_y}(\lambda_T)_2 \tag{7.186}$$

Now, we demand, that the scale of accretion corresponding to $\mathscr{T}$, namely

$$(\lambda_Y)_2 = \frac{\lambda_{\mathscr{T}}}{(\lambda_T)_2} = (\lambda_Y)_2 \lambda_{N_y} \tag{7.187}$$

is equal to the vertical model scale, i.e. that

$$\lambda_y = (\lambda_Y)_2 \,.\, \lambda_{N_y}$$

is valid. Substituting here the value of $(\lambda_Y)_2$ given by (7.184) we arrive at

$$\lambda_{N_y} = \lambda_y{}^{\frac{3}{2}} \tag{7.189}$$

Hence, the model erosion and/or accretion will be in the scale λ_y if each sequence of N_y' prototype (tidal) waves is simulated in the model by a sequence of

$$N_y'' = \lambda_y{}^{\frac{3}{2}} N_y'$$

(tidal) waves[18]. Observe that since λ_y is smaller than unity, N_y'' is always smaller than N_y'. In fact, one gets the impression that N_y'' perhaps is *too* small in comparison to N_y'. On the other hand, experiments confirm that the order of the ratio N_y'/N_y'' must be very large indeed. For example, in the case of the distorted model ($\lambda_y = 1/75$; $\lambda_x = 1/750$) of the Tees Estuary (H.R.S. Wallingford) (which reproduced accretion satisfactorily, as is clear from the diagrams in Fig. 7.15), the scale of accretion was nearest to the scale λ_y when the prototype year $\mathscr{T}'$ (consisting of $N_y' = 720$ tidal periods) was simulated in the model by the duration $\mathscr{T}'' = (T)_2$ corresponding to the passage of a *single* tide ($N_y'' = 1$). Considering that the number of tidal waves in the model should be an integer number, it follows that it is indeed the ratio

$$\frac{N_y''}{N_y'} = \frac{1}{720}$$

corresponding to the passage of the single tidal wave in the model that is nearest the value

$$\lambda_y^{\frac{3}{2}} = \frac{1}{75^{\frac{3}{2}}} = \frac{1}{632} \ 649$$

given by the equation (7.189). Using (7.182) and (7.189) in (7.186) we obtain

$$\lambda_{\mathscr{T}} = \frac{\lambda_y^2}{n} = \lambda_y \lambda_x \tag{7.190}$$

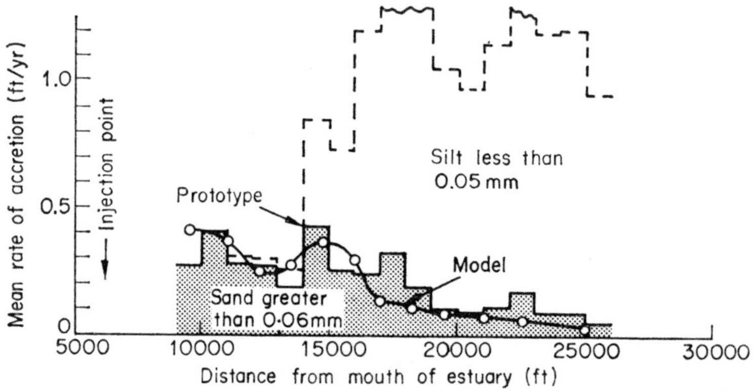

FIG. 7.15 The Tees Estuary
Comparison of siltation of course material in model and prototype

which coincides precisely with the value (6.78) of the corresponding time scale $(\lambda_t)_y$ for unidirectional flow. Hence, the model study of erosion and accretion due to the combination of unidirectional flow with the tidal flow is possible. One cannot be sure about the inclusion of short waves, for as has been pointed out earlier, one cannot be certain that the amount of grains lifted (eroded) by short waves in the model is in the required proportion.

(b) *Formation in the direction of x-axis*

From the derivation of eqn. (6.83) it is clear that this equation can be adopted also for bed material motion due to tidal waves. Introducing

$$\mathscr{T} = N_x(T)_2 \tag{7.191}$$

and thus

$$\lambda_{\mathcal{T}} = \lambda_{N_x}(\lambda_T)_2 \tag{7.192}$$

and identifying this value with $(\lambda_t)_x$ in eqn. (6.83) we obtain

$$\lambda_{N_x} = \frac{\sqrt{\lambda_y}}{n^{\frac{3}{2}}} \frac{1}{(\lambda_T)_2} \tag{7.193}$$

i.e.

$$\lambda_{N_x} = \frac{1}{\sqrt{n}} \tag{7.194}$$

Hence, the *nett* travel distances $\delta x''$ and $\delta x'$ of the model and prototype granular materials will be related to each other by the required proportion

$$\lambda_x = \frac{\delta x''}{\delta x'}$$

if each sequence of N_x' prototype (tidal) waves is simulated in the model by a sequence of

$$N_x'' = \frac{1}{\sqrt{n}} N_x'$$

(tidal) waves. Here again N_x'' is smaller than N_x' (but not so much as N_y'' is smaller in comparison with N_y').

In the case of the combined motion (consisting of short and long waves and the unidirectional flow), all three components have the same velocity scale $\sqrt{\lambda_y}$. Since the transport of granular material from one location to another is due to the flow velocities, it is obvious that model study of transport due to the combined motion is possible. Here the term 'transport' is being used in the sense of distance and time alone, in the travel of granular material. One cannot be sure with regard to the similarity of the *amount* of the transported material, if short waves are included.

Since λ_{Ny} and λ_{Nx} differ considerably, one cannot study erosion and transport during the same single run. They should be studied in two separate runs using their own different time scales.

EXAMPLE 7.4 ('Tauranga Harbour' model; HRS Wallingford)*

The investigations concern the improvement of the navigational and port facilities in Tauranga Harbour (New Zealand), and the changes in the mobile bed and sediment regime that would be likely to be produced in the inner harbour following the construction of the training wall, in particular.

* Hydraulics Research 1963–1966, Ministry of Technology, H.M. Stationery Office.

In addition, it is intended to examine the nature of the interchange of sediment between the outer entrance and the inner harbour.

From the above description, it is clear that the investigation is related directly to the transport of granular material and the formation of the mobile bed. Thus, these investigations must be carried out in a model with mobile bed. Let us consider the scales determining such a model.

The grain size of the prototype bed material cannot be regarded as large, and thus the method implied by $\lambda_{X_1} = \lambda_{X_2} = 1$ must be used. Furthermore, the action of short waves, long waves and littoral currents is to be considered, and thus the relations (7.166) of the combined motion must be applied.

The prototype order of $\bar{m}\beta$ is not known, and thus we can choose two scales freely, say λ_y and λ_x, insofar as the corresponding $\bar{m}\beta$ is acceptable. We select

$$\lambda_x = \frac{1}{840} \quad \text{and} \quad \lambda_y = \frac{1}{72}$$

which implies that the distortion is

$$n = 11{\cdot}65$$

while the value of $\bar{m}\beta$ is $\approx 1/4{\cdot}3$, which is a perfectly reasonable value (observe from Fig. 7.14 how near the point corresponding to the above values of λ_x and λ_y is to the shaded 'expected region'). Using the above values of λ_y and n in the relations (7.166) we obtain the following values for the scales determining the model:

$$\lambda_h = \frac{1}{72}$$

$$(\lambda_H)_1 = (\lambda_H)_2 = \frac{1}{72}$$

$$(\lambda_T)_1 = \sqrt{\frac{1}{72}} = \frac{1}{8{\cdot}5}$$

$$(\lambda_T)_2 = (\lambda_t)_3 = \sqrt{\left(\frac{1}{72}\right)}\,\frac{1}{11{\cdot}65} = \frac{1}{99}$$

$$\lambda_D = \frac{1}{\sqrt{\dfrac{11{\cdot}65}{72}}} = 2{\cdot}485 \approx 2{\cdot}5$$

$$\lambda_{\gamma_s} = \left(\frac{11{\cdot}65}{72}\right)^{\frac{3}{2}} \approx \frac{1}{15{\cdot}40}$$

Since $\gamma_s'/\gamma = 1{\cdot}65$ it follows that $\gamma_s''/\gamma = 1{\cdot}65/15{\cdot}40 = 0{\cdot}107$ and thus that the relative specific weight of the model bed material in air should be $\bar{\gamma}_s''/\gamma = 1{\cdot}107$. In the actual model shown in Plate 8 *coal* ($\bar{\gamma}_s''/\gamma = 1{\cdot}2$) was used as model bed material, the results were very satisfactory. The scales of some relevant properties of the Tauranga Harbour model are:

$$(\lambda_U)_1 = (\lambda_U)_2 = (\lambda_U)_3 = \sqrt{\frac{1}{72}} = \frac{1}{8{\cdot}5}$$

$$(\lambda_Q)_2 = (\lambda_Q)_3 = \frac{1}{8{\cdot}5} \times \frac{1}{840 \times 72} = \frac{1}{515\,000}$$

$$(\lambda_T)_2 = (\lambda_T)_3 = \frac{11{\cdot}65}{72} = \frac{1}{6{\cdot}18}$$

$$\lambda_p = \left(\frac{11{\cdot}65}{72}\right) \approx \frac{1}{15{\cdot}40} \qquad \text{(as implied by (6.67))}$$

$$\lambda_{\mathcal{T}} = \frac{1}{840} \times \frac{1}{72} = \frac{1}{60\,500}$$

Let N be the number of short waves in a long wave. According to (7.123) we have

$$\lambda_N = \frac{1}{11{\cdot}65} = \frac{1}{n}$$

(in order to ensure the equality of velocity scales).

Let N_y be the number of long waves in a typical prototype duration (say 1 year). The equality between the erosion scale λ_Y and the vertical model scale λ_y is ensured (according to (7.189)) by having

$$\lambda_{N_y} = \left(\frac{1}{72}\right)^{\frac{3}{2}} = \frac{1}{610}$$

The equality between the scale of the displacement (travel) of granular material along x and the horizontal model scale λ_x is ensured (according to (7.194)) by having

$$\lambda_{N_x} = \frac{1}{\sqrt{11{\cdot}65}} = \frac{1}{3{\cdot}42}$$

The last two relationships corresponding to two different runs.

EXAMPLE 7.5 ('Kuala Baram' model; HRS Wallingford)*

The Baram River (Sarawak) is navigable for a distance no less than 60 miles from its mouth. Yet the access to this river is obstructed by a sand bar over which the depth at low-water is as small as approximately four feet. The development of the hinterland is largely dependent upon the creation of a port capable of accepting ocean going ships, and thus on the possibility of having sufficiently large depths. It is intended to find ways of increasing the depths, and maintaining these depths, by model tests.

The investigation involves tides, short waves, littoral currents and salinity effects. The solution of the problem is directly related to the motion of the bed material, and thus a model with a mobile bed must be used. Here also the prototype grain size is small and thus the design of the model is entirely analogous to that of the previous example. The selected vertical and horizontal model scales are

$$\lambda_x = \frac{1}{600} \quad \text{and} \quad \lambda_y = \frac{1}{60}$$

which implies that

$$n = 10$$

and that the typical prototype value of $\bar{m}\beta$ is $\approx 1/4{\cdot}5$. The relationships (7.166) give the following values for the scales determining the model

$$\lambda_h = \frac{1}{60}$$

$$(\lambda_H)_1 = (\lambda_H)_2 = \frac{1}{60}$$

$$(\lambda_T)_1 = \sqrt{\frac{1}{60}} = \frac{1}{7{\cdot}75}$$

$$(\lambda_T)_2 = \sqrt{\frac{1}{60}} \times \frac{1}{10} = \frac{1}{77{\cdot}5}$$

$$\lambda_D = \sqrt{\frac{60}{10}} = 2{\cdot}45$$

$$\lambda_{\gamma_s} = \frac{1}{(2{\cdot}45)^3} = \frac{1}{14{\cdot}7}$$

* Hydraulics Research 1966, Ministry of Technology, H.M. Stationery Office London, 1967.

As for the previous example, we determine

$$(\lambda_U)_1 = (\lambda_U)_2 = (\lambda_U)_3 = \frac{1}{7\cdot75}$$

$$(\lambda_T)_2 = (\lambda_T)_3 = \frac{1}{6}$$

$$\lambda_N = \frac{1}{10} = \frac{1}{n}$$

$$\lambda_{N_y} = \left(\frac{1}{60}\right)^{\frac{3}{2}} = \frac{1}{465} \qquad \text{(for the formation along } y\text{)}$$

$$\lambda_{N_x} = \frac{1}{\sqrt{10}} = \frac{1}{3\cdot16} \qquad \text{(for the formation along } x\text{)}$$

An additional feature of this model is the fact that it does not operate with a simple homogeneous fluid. It is expected to operate with a salinity distribution similar to the prototype. Let ρ and $\bar{\rho}$ be the densities of the fresh and salt water at two (typical) locations A and $\bar{A}$ (where salinities are zero and maximim respectively). A correct distribution of the salinity in the intermediate regions can be achieved if the (additional) dimensionless variable

$$X_5 = \frac{\rho}{\bar{\rho}}$$

has identical values in model and prototype, i.e. if

$$\lambda_{X_5} = 1 \quad \text{or} \quad \lambda_\rho = \lambda_{\bar{\rho}}$$

is valid. Since the model operates with the same fluid as the prototype, it is clear that the condition above is equivalent to

$$\rho' = \rho'' \quad \text{and} \quad \bar{\rho}' = \bar{\rho}''$$

This condition is fulfilled in the Kuala Baram model.*

EXAMPLE 7.6 ('Karachi Harbour Model')[14]

This model is designed (for the government of Pakistan) by using the relation

$$\lambda_x = \lambda_y^{1\cdot43}$$

which corresponds to $\bar{m}\beta = 1/6$. The relation above is satisfied by selecting

$$\lambda_x = \frac{1}{200} \quad \text{and} \quad \lambda_y = \frac{1}{40}$$

and thus by having

$$n = 5$$

* For the reasons explained at the end of the section, it was eventually decided to use sand as granular material in the Kuala Baram model.

Hence, in comparison to the previous two examples the model scales are considerably larger while the distortion is smaller. In Ref. 14, short waves were not considered. However, the design can easily be generalized so as to include the short waves as well. Substituting $\lambda_y = 1/40$ and $n = 5$ in the scale relations (7.166) we obtain the following values

$$\lambda_h = \frac{1}{40}$$

$$(\lambda_H)_1 = (\lambda_H)_2 = \frac{1}{40}$$

$$(\lambda_T)_1 = \sqrt{\frac{1}{40}} = \frac{1}{6\cdot 32}$$

$$(\lambda_T)_2 = \sqrt{\left(\frac{1}{40}\right)\frac{1}{5}} = \frac{1}{31\cdot 6}$$

$$\lambda_D = \sqrt{\frac{40}{5}} = 2\cdot 83$$

$$\lambda_{\gamma_s} = \frac{1}{(2\cdot 83)^3} = \frac{1}{22\cdot 6}$$

the values of some relevant properties being

$$(\lambda_U)_1 = (\lambda_U)_2 = (\lambda_U)_3 = \frac{1}{6\cdot 32}$$

$$(\lambda_\tau)_2 = (\lambda_\tau)_3 = \frac{1}{8}$$

$$\lambda_N = \frac{1}{5} = \frac{1}{n}$$

$$\lambda_{N_y} = \left(\frac{1}{40}\right)^{\frac{3}{2}} = \frac{1}{253} \qquad \text{(for the formation along } y\text{)}$$

$$\lambda_{N_x} = \frac{1}{\sqrt{5}} = \frac{1}{2\cdot 24} \qquad \text{(for the formation along } x\text{)}$$

The required specific weight of the model bed material

$$\bar{\gamma}_s''/\gamma = \frac{1\cdot 65}{22\cdot 6} + 1 = 1\cdot 073$$

is noticeably less than in the preceding examples.

Before closing the present section it should be pointed out that in the case of wave motion (both wind and tidal waves), the velocity diagrams at the instant of change of flow direction, exhibit shapes shown schematically in Fig. 7.16. The larger the relative bed roughness k_s/h is (other dimensionless variables being equal), the less the curvature of the lower part of the diagram and the higher the position of the point where the velocity diagram changes sign[8].

Now, the method $\lambda_{X_1} = \lambda_{X_2} = 1$ considered here, which yields a model bed material lighter but coarser than the prototype material, yields at the

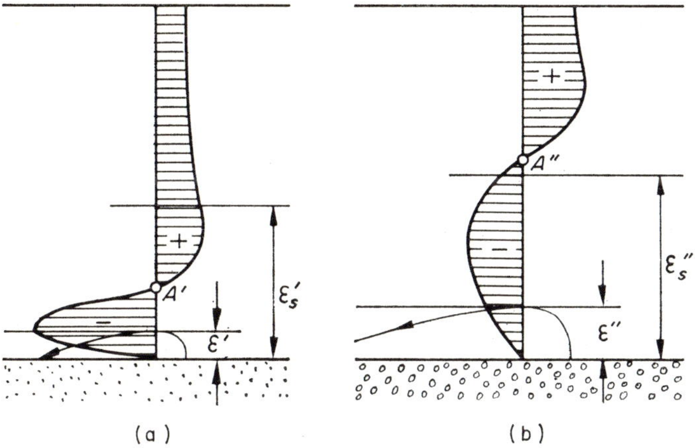

FIG. 7.16

same time the model value $(X_3'')^{-1} = D''/h''$, which is much larger than $(X_3')^{-1} = D'/h'$. Hence, the model designed according to $\lambda_{X_1} = \lambda_{X_3} = 1$ is much rougher than the prototype. So if the prototype velocity distribution is as in Fig. 7.16a, then the model distribution is as in Fig. 7.16b. The fact that the velocity distribution diagrams are not identical is only natural. Indeed, in the case of a distorted model, only one typical velocity can be reproduced correctly. For example, in a distorted model of a unidirectional flow, only the average velocity v (which was treated as a section property) is reproduced according to $\lambda_v = \sqrt{\lambda_y}$. The scales of the local velocities u corresponding to the homologous positions y/h are not equal to $\sqrt{\lambda_y}$ (see the expression $u/v_* = \varphi(y/k_s;\ v_* k_s/\nu)$). However, in the case of a unidirectional flow, all velocities u point in the same (downstream) direction, the sediment (at any level y) being always transported downstream. Hence, in a distorted model of a unidirectional flow, an error due to a non-similar distribution of the local velocities u is not so noticeable.

The same can be said for a model of two-directional flow (due to wave motion) if the sediment is transported in the vicinity of the bed (bed load); that is, if the dimensionless excess $Y - Y_{cr}$ of the tractive force is small and the 'ceilings' ε' and ε'' of the regions occupied by the transported materials are lower than A' and A'' respectively (Fig. 7.16a and b).* Conversely, if, the value of $Y - Y_{cr}$ is large, and the transport of granular material is not confined to the vicinity of the bed (appreciable suspended load), then the error due to non-similarity in the distribution of local velocities can be very noticeable. Indeed, if, for example, the regions occupied by the prototype and model materials transported in suspension are ε_s' and ε_s'', then although the prototype transport might occur in comparable rates in both directions, the model can predict that it occurs totally in only one direction. Clearly this kind of prediction is unacceptable, and therefore when transport in suspension is appreciable, one frequently departs from the method $\lambda_{X_1} = \lambda_{X_2} = 1$ and prefers to use sand (rather than, say, perspex or polystyrene) as the model bed material. By doing so, a certain error is introduced with regard to the similarity of the threshold (initiation of transport), and also with regard to the reproduction of the transport rate. This is because the scale λ_p can be given only if the property Π_p is a function of X_1 and X_2 and if the model and prototype values of these variables are identical. On the other hand, by using sand one can have a smaller relative bed roughness, a more realistic distribution of local velocities, and consequently a more reliable set of conditions as far as judgment on the transport directions and their levels is concerned.**

7.9 Large Values of the Reynolds Number X_1'

If the prototype grain size D_{50}', or to be more precise, if the prototype Reynolds number

$$X_1' = \frac{v_*' D_{50}'}{\nu}$$

is large, then its influence can be ignored. In this case, the dynamic similarity of the two phase motion can be given by

$$\lambda_{X_2} = 1 \quad \lambda_{X_3} = 1 \tag{7.195}$$

* In the case of the bed load the grains are moving by discrete jumps, the height ε being the height of the jump; the ratio $\varepsilon''/\varepsilon'$ being equal to λ_D.

** This relevant restriction of the $\lambda_{X_1} = \lambda_{X_2} = 1$ method has been brought to the author's attention by R. C. H. Russell.

Assuming that the model fluid is water, these conditions give

$$\left.\begin{aligned} \lambda_U &= \sqrt{(\lambda_{\gamma_s}\lambda_D)} \\ \lambda_l &= \lambda_D \\ \lambda_{\delta_s} &= 1 \end{aligned}\right\} \qquad (7.196)$$

Here U and l are velocity and length, in general, while the equations (7.196) are valid for short waves, long waves, unidirectional current, or for any of their combinations. Thus U can be interpreted as U_δ, v_*, v, etc., while l can be h, a_δ, H, and so on. Indeed, we assume that the model is undistorted, and thus that

$$\lambda_l = \lambda \qquad (7.197)$$

is valid whatever the interpretation of l might be. Furthermore we assume that this undistorted model is Froudian, and consequently that λ_U given by

$$\lambda_U = \sqrt{\lambda} \qquad (7.198)$$

is the scale of any velocity.

Let us show that the relations (7.197) and (7.198) are compatible with the mobile bed conditions (7.196). The condition $\lambda_{\rho_s} = 1$ together with $\lambda_g = \lambda_\rho = 1$ yields at once

$$\lambda_{\gamma_s} = 1 \qquad (7.199)$$

But if so, then the first two equations of (7.196), in fact, are simply (7.197) and (7.198) provided that

$$\lambda_D = \lambda \qquad (7.200)$$

is valid. Hence, the dynamic similarity of two phase motion (corresponding to large X_1') in a small scale undistorted Froudian model can be achieved

(a) if the specific weight of the model grains is identical to those of the prototype (i.e. if the model bed material is sand or gravel), and
(b) if the grain size is reduced in the same proportion λ as the dimensions of flow.

Observe that since the condition $X_4' = X_4''$ is satisfied, the dynamic similarity includes both similarity in the behaviour of bed material *en mass* as well as similarity in the behaviour of individual grains. The model scale λ of the undistorted model under consideration must satisfy the condition analogous to (5.16) introduced in Chapter 5.

EXAMPLE 7.7 (Seaford Sea Wall Model; HRS Wallingford)[19]
It is intended to determine by model tests the quantity of pebbles A that must be placed on the foreshore to ensure that the ground B and consequently the foundations C of the sea wall are adequately protected

(Fig. 7.17). Typical sizes of pebbles are: $D_{10}' = 14$ mm; $D_{50}' = 24$ mm; $D_{50}' = 49$ mm, the porosity being $n' = 0{\cdot}375$. The significant wave height $H' = 4$ m ($\approx$12 ft); the wave period $T' = 10$ sec. It is assumed that the duration of the storm is 12 hours. The scale of the undistorted model is $\lambda = 1/20$.

The velocities U' generated by the waves and the size of the pebbles D' are sufficiently large to enable any influence of the Reynolds number $U'D'/\nu$ to be ignored so far as the motion of pebbles due to wave motion is concerned. However, the flow that percolates through the grains, with

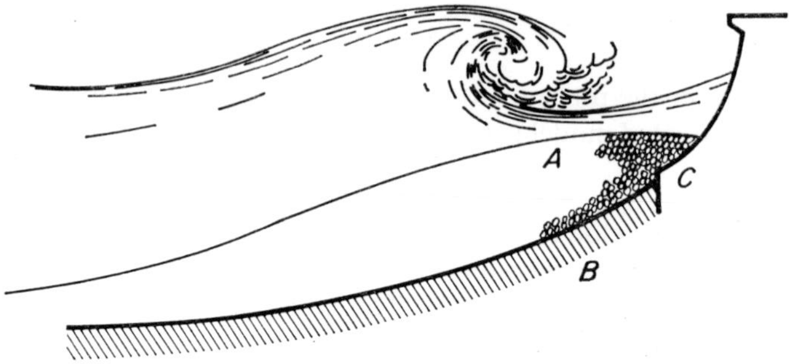

FIG. 7.17

the velocity V' is not completely independent of viscosity. Since the amount of fluid percolating in, and moving through, the porous pebble-medium (during the passage of each wave) has a certain influence on the mechanical behaviour of the water-pebble system, while the influence of the pebble roughness on the wave motion is negligible, the grain size scale λ_D has been chosen to yield similarity in percolating flow, rather than $\lambda_D = \lambda$. Accordingly, the model has been designed to have the following dimensionless quantities the same as the prototype

$$\mathrm{Fr} = \frac{U^2}{gl} \quad \text{Froude number}$$

$$Y_U = \frac{\rho U^2}{\gamma_s D} \quad \text{Mobility number of grains}$$

$$\frac{U}{V} = \frac{\text{wave velocity}}{\text{filtration velocity}}$$

$$J = f(\mathrm{Re})\frac{V^2}{gD} \quad \text{Filtration energy gradient}$$

$$\left(\text{with } \mathrm{Re} = \frac{VD}{\nu}\right)$$

The identity of the quantities above yields the scale relations

$$\lambda_U{}^2 = \lambda_v{}^2 = \lambda$$
$$\lambda_{\gamma_s}\lambda_D = \lambda$$
$$\lambda_f = \frac{\lambda_D}{\lambda}$$

where

$$\lambda_f = \frac{f(\mathrm{Re}'')}{f(\mathrm{Re}')} = \frac{f(\lambda_v\lambda_D\mathrm{Re}')}{f(\mathrm{Re}')}$$

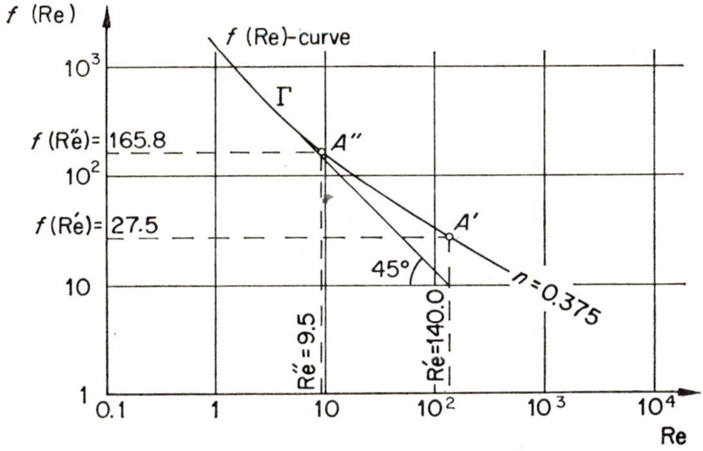

FIG. 7.18

Hence

$$\lambda_U = \lambda_v = \sqrt{\lambda} = \sqrt{\frac{1}{20}} = \frac{1}{4{\cdot}47}$$

whereas λ_{γ_s} and λ_D are given by

$$\lambda_{\gamma_s}\lambda_D = \frac{1}{20}$$

and

$$\frac{f\left(\dfrac{\lambda_D}{4{\cdot}47}\,140\right)}{f(140)} = 20\lambda_D$$

where 140 is the estimated value for the prototype Reynolds number Re′. The curve $f(\mathrm{Re})$ corresponding to the porosity $n = 0{\cdot}375$ is shown in Fig. 7.18. Here the prototype point A' is known. It is the point which corresponds to $\mathrm{Re}' = 140$. The model point A'' is not known. It must be determined by using the relation above which can be written as

$$f(31{\cdot}4\lambda_D) = 20\lambda_D f(140) = 20 \times 27{\cdot}5\lambda_D$$

i.e.

$$f(31{\cdot}4\lambda_D) = 550\lambda_D$$

Now the value of λ_D must be selected (by trial and error) so that the point A'' having the abscissa

$$\mathrm{Re}'' = 31{\cdot}4\lambda_D$$

and the ordinate

$$f(\mathrm{Re}'') = 550\lambda_D$$

occurs on the curve $f(\mathrm{Re})$. This is possible for *one* value of λ_D only. This value is $\lambda_D = 1/3{\cdot}3$ and it gives

$$\mathrm{Re}'' = 31{\cdot}4\,\frac{1}{3{\cdot}3} = 9{\cdot}5$$

and

$$f(\mathrm{Re}'') = 550\,\frac{1}{3{\cdot}3} = 165{\cdot}8$$

Hence the size of the geometrically similar 'model pebble' must be 3·3 times smaller than that of the prototype, the specific weight scale being

$$\lambda_{\gamma_s} = \frac{1}{20\lambda_D} = \frac{3{\cdot}3}{20} = 0{\cdot}165$$

which gives

$$\bar{\gamma}_s''/\gamma = 1 + 0{\cdot}165\,.\,1{\cdot}65 = 1{\cdot}272$$

A light weight coal ($\bar{\gamma}_s''/\gamma = 1{\cdot}27$) was chosen as the 'model pebble'.

REFERENCES

1. *A summary of the theory of oscillatory waves*, Beach Erosion Board, Techn. Rep. No. **2**, U.S. Government Printing Office (1942).
2. H. Lamb, *Hydrodynamics*, Dover Publications (1945).
3. L. M. Milne-Thomson, *Theoretical Hydrodynamics*, Chapter XV, Macmillan (1968).
4. N. J. Kochin, I. A. Kibel, N. W. Rose, *Theoretical Hydromechanics*, Vol. **I**, Chapter VIII, John Wiley (1965).
5. P. S. Eagleson, R. G. Dean, *Small Amplitude Wave Theory*, Chapter 1, in *Estuary and Coastline Hydrodynamics* (Ed. A. T. Ippen), McGraw-Hill (1966).

6. H. Schlichting, *Boundary Layer Theory*, Chapter 5, McGraw-Hill (6th edition) (1968).
7. H. Ludwig, W. Tillman, *Investigation of the Wall Shearing Stress in Turbulent Boundary Layers*, NACA Tech. Memo. No. 1285 (1950).
8. M. S. Yalin, R. C. H. Russell, 'Shear Stress Due to Long Waves', *Journal of Hydraulics Research*, **4,** No. 2 (1966).
9. G. H. Keulegan, *Model Laws for Coastal and Estuarine Models*, Ch. 17 *Estuary and Coastline Hydrodynamics* (Ed. A. T. Ippen), McGraw-Hill (1966).
10. I. G. Jonsson, *Measurements in the Turbulent Wave Boundary Layer*, Tech. Univ. Denmark, Coastal Eng. Lab., Basic Res. Prog. Rep. No. 5 (1964).
11. I. G. Jonsson, *On the Existence of Universal Velocity Distributions in an Oscillatory, Turbulent Boundary Layer*, Tech. Univ. Denmark, Coastal Eng. Lab., Basic Res. Prog. Rep. No. **12** (1966).
12. H. Li, *Stability of Oscillatory Laminar Flow Along a Wall*. Beach Erosion Board Tech. Memorandum. No. **47,** Washington (1954).
13. R. C. H. Russell, M. S. Yalin, *Similarity in Sediment Transport Due to Waves*, Proc. 8th Conf. on Coastal Engineering, Chapter XII, Mexico City (1962).
14. S. Z. H. Rizvi, *Similarity Laws for Hydraulic Models with Mobile Beds*, D.S.I.R., H.R.S. Wallingford, Rep. No. INT. 30 (1963).
15. E. W. Bijker, *Determination of Scales of Moveable Bed Models*, Delft, Hydraulics Lab. Publication No. **35,** May (1965).
16. H. C. Frijlink, 'Discussion des formules de débit solide de Kalinske, Einstein et Meyer-Peter et Müller,' *Comptes rendus Zième Journee Hydr., Soc. Hydrotechn. de France*, Grenoble (1952).
17. M. S. Longuet-Higgins, *The Mechanics of the Boundary Layer Near the Bottom in a Progressive Wave*. Proc. 6th Conf. on Coastal Engineering, Chapter X–Appendix (1957).
18. M. S. Yalin, *Method for Selecting Scales for Models with Movable Bed Involving Wave Motion and Tidal Currents*, Proc. 10th Cong., I.A.H.R., London (1963).
19. M. S. Yalin, *A Model Shingle Beach with Permeability and Drag Forces Reproduced*, Proc. 10th Cong. I.A.H.R., London (1963).
20. M. S. Yalin, 'On the Average Velocity of Flow over a Mobile Bed', *La Houille Blanche*, No. 1, Jan.–Feb. (1964).

List of Notations

(i) Theory of Dimensions and Models

L	Length unit
T	time unit
M	mass unit
a, a_i	characteristic parameter (of a phenomenon)
X, X_j	dimensionless variable (of a phenomenon)
A	dimensional property (of a phenomenon)
Π_A	dimensionless version of the property A
f_A	dimensional function (determining A)
φ_A	dimensionless function (determining Π_A)
a', (A')	prototype value of a (of A)
a'', (A'')	model value of a (of A)
$\lambda_a = \frac{a''}{a'} \left(\lambda_A = \frac{A''}{A'}\right)$	the scale of a (of A)
λ_x	horizontal model scale
λ_y	vertical model scale
$n = \lambda_y/\lambda_x$	distortion
λ	geometric scale of an undistorted model
Re	Reynolds Number
Fr	Froude Number
St	Strouhal Number
Ma	Mach Number
l	typical length
U	typical velocity

(ii) Physical Properties

g	acceleration due to gravity	
ρ	fluid density	fluid
μ	dynamic viscosity	fluid
ν	kinematic viscosity	fluid
$\gamma = \rho g$	specific weight of fluid	fluid
ρ_s	density of grains	
$\bar{\gamma}_s = \rho_s g$	specific weight of grains (in air)	granular material
$\gamma_s = \bar{\gamma}_s - \gamma$	specific weight of grains (in fluid)	granular material
D	grain size (in general)	granular material
$\bar{D}$	effective grain size	granular material
D_p	grain size corresponding to the percentage p (of the mixture)	granular material
k	permeability coefficient	granular material
κ	Von Karman constant	

(iii) Coordinates

x, y, z	ortagonal space coordinates (the direction of x coincides with the direction of the flow, y being in the vertical plane. Usually y is the distance from the flow boundary upwards)
ξ, η, ζ	dimensionless versions of x, y, z
t	time
θ	dimensionless time

(iv) Quantities Related to the Motion of Fluid (and of granular material)

v	average velocity of the flow
u	local velocity (at the distance y from the flow boundary)
τ	local shear stress (at the distance y from the flow boundary)
τ_0	shear stress interacting between fluid and the flow boundary (at $y = 0$)
$v_* = \sqrt{\tau_0/\rho}$	shear velocity
δ	thickness of the boundary layer
$\bar{\delta}$	thickness of the viscous sublayer (of a turbulent flow)
h	flow depth
B	the width of the open channel flow at the free surface
b	the width of the channel bed
B_c	the width of the (central) two dimensional region of the open channel flow
ω	flow cross-section

χ	wetted perimeter	
$R = \omega/\chi$	hydraulic radius	
k	size of the flow boundary roughness	
k_s	size of the (equivalent) sand roughness	
$c = v/v_*$	dimensionless friction factor	
$\lambda = 8/c^2$	Weisbach (friction) coefficient	
$C = \sqrt{g}c$	Chezy (friction) coefficient	
H	energy head	
$J = -\dfrac{\partial H}{\partial x}$	energy gradient (along x)	
ζ_l	local energy loss coefficient	
$E = J/\mathrm{Fr}$	energy loss coefficient	
S	slope of the channel bed	
$\bar{S}$	slope of the free surface	
$Q = v\,.\,\omega$	flow rate (flow discharge)	
q	specific flow rate (per unit width of the flow)	
L	wave length	wave motion
H	wave height	wave motion
T	wave period	wave motion
a	horizontal orbit length	wave motion
$U = 2a/T$	mean orbital velocity (at the bed)	wave motion
P	total sediment transport rate (weight per unit time)	motion of granular material
p	specific sediment transport rate (weight per unit time and width)	motion of granular material
Λ	sand wave length	motion of granular material
Δ	sand wave height	motion of granular material
Y	elevation of the surface of the mobile bed	motion of granular material

Author Index

Subject Index